SCHOLAR Study Guide
Advanced Higher Biology

Authored by:
Bryony Clutton (North Berwick High School)
Dawn Campbell (Falkirk High School)

Reviewed by:
Fiona Stewart (Perth Grammar School)
Nicola Haddow (Levenmouth Academy)

Previously authored by:
Eileen Humphrey
Jaquie Burt
Lorraine Knight
Nadine Randle

Heriot-Watt University
Edinburgh EH14 4AS, United Kingdom.

First published 2019 by Heriot-Watt University.

This edition published in 2019 by Heriot-Watt University SCHOLAR.

Copyright © 2019 SCHOLAR Forum.

Members of the SCHOLAR Forum may reproduce this publication in whole or in part for educational purposes within their establishment providing that no profit accrues at any stage, Any other use of the materials is governed by the general copyright statement that follows.

All rights reserved. No part of this publication may be reproduced, stored in a retrieval system or transmitted in any form or by any means, without written permission from the publisher.

Heriot-Watt University accepts no responsibility or liability whatsoever with regard to the information contained in this study guide.

Distributed by the SCHOLAR Forum.

SCHOLAR Study Guide Advanced Higher Biology

Advanced Higher Biology Course Code: C807 77

ISBN 978-1-911057-66-6

Print Production and Fulfilment in UK by Print Trail www.printtrail.com

Acknowledgements

Thanks are due to the members of Heriot-Watt University's SCHOLAR team who planned and created these materials, and to the many colleagues who reviewed the content.

We would like to acknowledge the assistance of the education authorities, colleges, teachers and students who contributed to the SCHOLAR programme and who evaluated these materials.

Grateful acknowledgement is made for permission to use the following material in the SCHOLAR programme:

The Scottish Qualifications Authority for permission to use Past Papers assessments.

The Scottish Government for financial support.

The content of this Study Guide is aligned to the Scottish Qualifications Authority (SQA) curriculum.

All brand names, product names, logos and related devices are used for identification purposes only and are trademarks, registered trademarks or service marks of their respective holders.

Contents

1 Cells and Proteins — 1

1 Laboratory techniques for biologists 3
2 Proteomics, protein structure, binding and conformational change 37
3 Membrane proteins 67
4 Communication within multicellular organisms 83
5 Protein control of cell division 113
6 Cells and proteins test 131

2 Organisms and Evolution — 143

1 Field techniques for biologists 145
2 Evolution 155
3 Variation and sexual reproduction 163
4 Sex and behaviour 181
5 Parasitism 195
6 Organisms and evolution test 229

3 Investigative Biology — 243

1 Scientific principles and process 245
2 Experimentation 257
3 Critical evaluation of biological research 285
4 Investigative biology test 299

Glossary — 308

Answers to questions and activities — 317

Cells and Proteins

1	**Laboratory techniques for biologists**	**3**
1.1	Health and safety	6
1.2	Liquids and solutions	8
1.3	Separation techniques	12
1.4	Detecting proteins using antibodies	18
1.5	Microscopy	21
1.6	Aseptic technique and cell culture	24
1.7	Learning points	31
1.8	End of topic test	33
2	**Proteomics, protein structure, binding and conformational change**	**37**
2.1	The proteome	41
2.2	The synthesis and transport of proteins	42
2.3	Protein structure	49
2.4	Ligand binding	57
2.5	Reversible binding of phosphate and the control of conformation	59
2.6	Learning points	61
2.7	Extended response questions	63
2.8	End of topic test	63
3	**Membrane proteins**	**67**
3.1	Movement of molecules across membranes	69
3.2	Ion transport pumps and generation of ion gradients	75
3.3	Learning points	77
3.4	Extended response question	79
3.5	End of topic test	80
4	**Communication within multicellular organisms**	**83**
4.1	Coordination	86
4.2	Hydrophobic signals and control of transcription	87
4.3	Hydrophilic signals and transduction	91
4.4	Generation of a nerve impulse	96
4.5	Initiation of a nerve impulse in response to an environmental stimulus: the vertebrate eye	103

4.6	Learning points	106
4.7	Extended response question	109
4.8	End of topic test	110

5 Protein control of cell division **113**

5.1	The cytoskeleton and cell division	115
5.2	The cell cycle	117
5.3	Control of the cell cycle	119
5.4	Control of programmed cell death (apoptosis)	122
5.5	Learning points	125
5.6	Extended response questions	127
5.7	End of topic test	128

6 Cells and proteins test **131**

Unit 1 Topic 1

Laboratory techniques for biologists

Contents

1.1 Health and safety 6
1.2 Liquids and solutions 8
1.3 Separation techniques 12
1.4 Detecting proteins using antibodies 18
1.5 Microscopy 21
1.6 Aseptic technique and cell culture 24
1.7 Learning points 31
1.8 End of topic test 33

Prerequisites

You should already know that:

- when culturing micro-organisms, their growth media requires raw materials for biosynthesis as well as an energy source;

- culture conditions include sterility (to eliminate any effects of contaminating microorganisms), control of temperature, control of oxygen levels by aeration, and control of pH by buffers or the addition of acid or alkali.

UNIT 1. CELLS AND PROTEINS

Learning objective

By the end of this topic, you should be able to:

- explain how to identify and control hazards and assess risk in a lab environment;
- state that substances, organisms and equipment in a laboratory can present a hazard;
- state that hazards in the lab include toxic or corrosive chemicals, heat or flammable substances, pathogenic organisms and mechanical equipment;
- define the terms 'risk' and 'risk assessment' in relation to health and safety;
- describe the use of control measures, including using appropriate handling techniques, protective clothing and equipment, and aseptic technique;
- describe how to carry out linear and log dilution series;
- describe how unknown concentrations can be determined using a standard curve;
- describe the role of buffers in maintaining and controlling pH;
- describe the method and uses of a colorimeter to quantify concentration and turbidity;
- describe the use of centrifugation to separate substances of differing density;
- describe the use of paper and thin layer chromatography for separating different substances, e.g. amino acids and sugars;
- describe the principle of affinity chromatography and its use in separating proteins;
- describe the principle of gel electrophoresis and its use in separating proteins and nucleic acids;
- describe the difference between native gels and SDS PAGE;
- state that proteins can be separated from a mixture using their isoelectric points (IEPs);
- state that IEP is the pH at which a soluble protein has no net charge and will precipitate out of solution;
- state that If the solution is buffered to a specific pH, only the protein(s) that have an IEP of that pH will precipitate;
- describe how proteins can also be separated using their IEPs in electrophoresis;
- state that immunoassay techniques are used to detect and identify specific proteins;
- state that these techniques use stocks of antibodies with the same specificity, known as monoclonal antibodies;
- describe the types of labels used to identify antibody binding;
- describe the use of western blotting;
- state that bright field microscopy can be used to examine whole organisms, parts of organisms, thin sections of dissected tissue or individual cells;

TOPIC 1. LABORATORY TECHNIQUES FOR BIOLOGISTS

Learning objective continued

- state that fluorescence microscopy uses specific fluorescent labels to bind to and visualise certain molecules or structures within cells or tissues and allows particular protein structures to be visualised;

- define the term 'aseptic techniques' and give examples;

- state that a microbial culture can be started using an inoculum of microbial cells on an agar medium or in a broth with suitable nutrients;

- describe the role of growth factors from serum in culturing animal cells;

- state that in culture, primary cell lines can divide a limited number of times, whereas tumour cells lines can perform unlimited divisions;

- state that plating out of a liquid microbial culture on solid media allows the number of colony-forming units to be counted and the density of cells in the culture estimated;

- state that serial dilution is often needed to achieve a suitable colony count;

- describe the method and use of haemocytometer to estimate cell numbers in a liquid culture;

- describe the use of vital staining to identify and count viable cells.

6 UNIT 1. CELLS AND PROTEINS

1.1 Health and safety

You will be familiar with health and safety rules from working in a school laboratory. Many biological laboratory health and safety rules are similar to those in a school laboratory and most are common sense. For example, safety glasses/goggles should be worn at all times while working with chemicals, heat or glassware to protect the eyes from potential harm.

The following table details some health and safety issues which have arisen in research laboratories and the recommended health and safety actions which should be put in place to minimise potential harm.

Concern statement	Health and safety recommended actions
Two scientists working in a laboratory experienced health and safety issues relating to liquid nitrogen. The main hazard related to working with liquid nitrogen is extremely low temperatures (in the order of -200°C or lower), which can lead to tissue damage. The first incident involved an employee filling a four litre bottle with liquid nitrogen. After stopping the flow of liquid nitrogen to check how full the bottle was, the researcher opened the liquid flow valve to continue filling. Restarting the flow of liquid nitrogen resulted in liquid nitrogen splashing out of the bottle and onto her hand. In a second case, a researcher received first and second degree contact burns to his left hand when he tried to shut the liquid flow valve on a 200 litre liquid nitrogen bottle.	Researchers and staff should be reminded of the importance of appropriate personal protective equipment. For example, when handling liquid nitrogen, appropriate eye protection includes safety glasses with side shields (if working with small volumes of liquid nitrogen), and safety glasses with side shields and a face shield (if working with liquid nitrogen from a pressurised line). Insulated gloves should be worn at all times to protect against extremely low temperatures. Good lab practice, such as wearing closed-toe shoes, trousers (not shorts or skirts) and a lab coat, is also required.
A centrifuge spins samples at high speed to separate out the components. A lab was running a centrifuge containing milk samples, which were placed in the rotor section. The rotor holds individual sample tubes and is connected to the spin drive of the centrifuge. Halfway through the procedure, the rotor failed due to excessive mechanical stress; this caused an explosion which destroyed the centrifuge. The safety shielding on the centrifuge failed and fragments of the centrifuge, including the steel rotor, were ejected from the machine. Metal fragments made holes in the walls and ceilings, and four windows were shattered due to the shockwave created by the explosion. Although the room was severely damaged, it was unoccupied at the time of the explosion and no injuries were reported. The explosion was caused by the use of an incorrect rotor in the centrifuge.	The rotor in a centrifuge is spun at extremely high speed and this causes powerful mechanical stress that can cause the rotor to fail. A centrifuge must be loaded correctly or the rotor may break loose while spinning. All researchers using a centrifuge must be aware of the proper operating procedures, including how to select, load, balance and clean the rotor. Laboratory supervisors are responsible for ensuring that all researchers are properly trained and that equipment is well maintained. If equipment is checked and there are signs of wear or damage to rotors, the equipment should be taken out of service immediately and clearly marked 'Warning - do not use' until checked by an authorised service technician.

Concern statement	Health and safety recommended actions
Recently, in a research lab, five persons required medical treatment for cuts sustained from broken glassware within a six month period.	All laboratories must have a formal plan for handling glassware, which should be stored in a safe place. Safe working areas should be provided for all work with glassware, i.e. glassware washing areas should have lots of clear space, sufficient safe drying racks and safe glassware storage areas. Appropriate gloves should be worn, i.e. slip resistant. Glassware should be regularly inspected, and broken, cracked or chipped items should be discarded. All researchers should be aware of proper procedures for disposal of broken glass.
Whilst operating a UV transilluminator, a researcher failed to wear the appropriate personal protective equipment (PPE). After using the transilluminator multiple times over a one hour period, the researcher suffered reddening of the skin and temporary eye injury resulting from UV exposure to her face and eyes.	A UV transilluminator emits UV light at intensities many times greater than the summer midday sun. Therefore, it can cause severe damage to the skin and eyes, even when standing several feet away. The light box shield must be kept in place at all times. Appropriate personal protective equipment must be used when working with a UV light source, including a face shield, safety glasses, gloves and a fully buttoned lab coat. All researchers must be made aware of the safe use of a UV transilluminator.

Chemicals or organisms can be intrinsically hazardous. Their use may involve risks to people, to other organisms or to the environment. Risk is the likelihood of harm arising from exposure to a hazard. A risk assessment involves identifying control measures to minimise the risk. Risk assessments must be performed when using certain substances in a laboratory. Control of substances hazardous to health (COSHH) regulations cover substances that are hazardous to health, as the name implies, including:

- chemicals;
- products containing chemicals;
- fumes;
- dusts;
- vapours and mists;
- nanotechnology;
- gases and asphyxiating gases;
- biological agents.

A COSHH assessment form allows a risk assessment to be carried out for any substance which is potentially hazardous. In this form, factors such as exposure, disposal procedures and handling procedures are considered. Follow the link to view a COSHH assessment form:
http://www.hse.gov.uk/nanotechnology/coss-assessment-form.doc

One of the most important aspects of health and safety is personal protective equipment (PPE), which reduces the risk of hazardous materials coming into contact with the skin where it could cause harm.

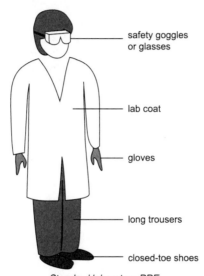

Standard laboratory PPE

Many biology laboratories work with living organisms. Living organisms can pose potential harm to the health and safety of people working in the laboratory, and the environment, therefore procedures must be put in place to minimise risk. For example, researchers working with microorganisms must follow stringent guidelines to ensure not only their own health and safety, but also that of others working in the laboratory, whilst additionally protecting the environment.

1.2 Liquids and solutions

Dilution series

Serial dilutions are an important laboratory technique. Dilution involves reducing the concentration of a substance in a solution. Repeated dilution from a stock solution is known as a serial dilution. Dilutions may be performed on a linear or a log scale. Dilutions in a linear dilution series differ by an equal interval, for example 0·1, 0·2, 0·3 and so on. Dilutions in a log dilution series differ by a constant proportion, for example 10^{-1}, 10^{-2}, 10^{-3} and so on.

The following diagram demonstrates how a serial dilution can be performed. 1 cm^3 of the stock solution is transferred to a new tube and 9 cm^3 of water is added to produce a concentration of 10^{-1}.

To produce a concentration of 10^{-2}, 1 cm³ of the 10^{-1} solution is removed and placed in a new tube and 9 cm³ of water is added. This process is repeated until the lowest required concentration has been produced. Each step is a tenfold dilution; therefore, this is an example of a logarithmic dilution.

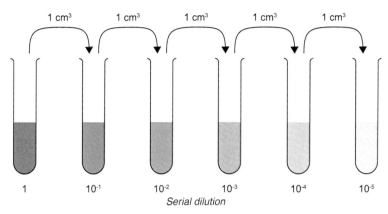

Serial dilution

Serial dilutions are widely performed in microbiology. For example, an investigator may need to know the number of bacterial cells which are contained within an environmental sample. It is likely that there are too many cells in the sample to physically count every one; therefore, we can carry out a dilution series. A dilution series allows the scientist to dilute the original concentration of bacteria in a stepwise manner. A sample from each dilution can be added to an agar plate and incubated until colonies develop. The investigator will select a plate with a countable number of colonies (usually between 30 and 300). By counting the number of colonies on a plate and knowing the concentration of the solution, the investigator can assess the number of cells per cm³ in the original culture.

Colorimeter

The concentration of a pigmented compound can be quantified using a **colorimeter**. A colorimeter measures the absorbance of specific wavelengths of light by a solution. A colorimeter works by passing a light beam, at a specific wavelength, through a cuvette containing a sample solution. Some of the light is absorbed by the sample; therefore, light of a lower intensity hits the detector and the machine will display an absorbance or transmission value.

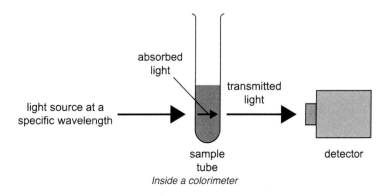

Inside a colorimeter

To use a colorimeter, the correct wavelength of light must first be selected (directions for most colorimetry experiments usually give a recommended wavelength) and the machine must be calibrated. To calibrate a colorimeter, a cuvette containing calibration solution (usually distilled water) is placed in the cuvette compartment and the calibrate button is pressed. This will provide the user with a reading of 0 absorbance. The user can then place a cuvette containing a sample solution into the cuvette compartment and take a reading of the absorbance of the sample by pressing the test button.

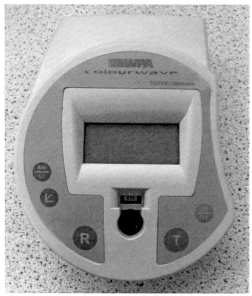

A colorimeter

Colorimeters can give two types of readings:

1. absorbance - how much light has been absorbed by the sample;
2. transmission - how much light passed through the sample without being absorbed.

Absorbance can be used to determine concentration of a coloured solution using suitable wavelength filters. Percentage transmission can be used to determine turbidity, such as cells in suspension (as the concentration of cells increases, the turbidity of the culture increases).

It can often be important to determine unknown concentrations of solutions in a laboratory. A **standard curve** is one method that is used to determine the concentration of a solution. A series of 'standards' of known concentration are measured and graphed. This graph can then be used to determine the concentration of an unknown sample.

One such example is the Bradford protein assay. In this test, the depth of colour produced by coomassie brilliant blue dye changes depending on the concentration of protein in the sample; this can be measured using a colorimeter.

To produce a standard curve, coomassie brilliant blue dye is added to known concentrations of protein solutions. The ideal protein solution to use is a purified preparation of the unknown sample; however, this may not always be available. In the absence of purified preparations of the protein being analysed, protein standards may be used. The two most common protein standards used are BSA and gamma-globulin. The absorbance of each known protein solution is measured using a colorimeter with a filter at 595 nm. The absorbance of each solution can then be plotted on a graph with the concentration on one axis and the absorbance on the other, producing a standard curve. The same dye (coomassie brilliant blue) is added to the protein sample of unknown concentration and an absorbance reading is taken. The standard curve can then be used to determine the concentration of protein in the sample based on its absorbance.

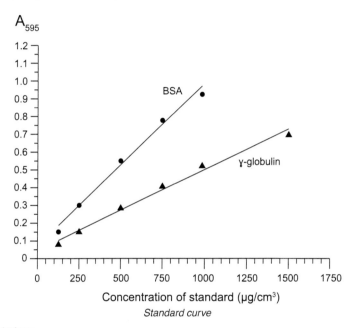

Standard curve

pH of solutions

pH is a measure of the acidity or basicity of a solution. Acidic solutions have a pH of less than 7 and basic (or alkaline) solutions have a pH of greater than 7. A solution with a pH of 7 is said to be neutral. pH is a logarithmic measure of hydrogen ion concentration.

$$pH = -\log\left[H^+\right]$$

The pH of a solution can be measured using an electronic meter or an indicator.

A pH buffer is a solution whose pH changes very little when a small amount of acid or base is added to it. Buffers work by allowing the addition of hydrogen or hydroxide ions without affecting the pH of the solution. Buffer solutions are used as a means of keeping pH at a nearly constant value. A buffer of carbonic acid and bicarbonate is present in blood plasma to maintain a pH between 7.35 and 7.45.

1.3 Separation techniques

Substances can be separated according to their solubility, size, shape or charge. The techniques we will look at are:

- **centrifugation**;
- gel **electrophoresis**;
- iso-electric point;
- paper, thin layer and affinity **chromatography**.

Centrifugation

A centrifuge is a piece of equipment that spins a sample at high speed.

A benchtop centrifuge by http://commons.wikimedia.org/wiki/User:Magnus_Manske, licenced under the Creative Commons
http://creativecommons.org/licenses/by/1.0/deed.en license

Centrifugation allows substances to be separated according to their density. The densest materials separate out first and form a pellet at the bottom of the tube. The liquid which remains above the pellet is called the supernatant.

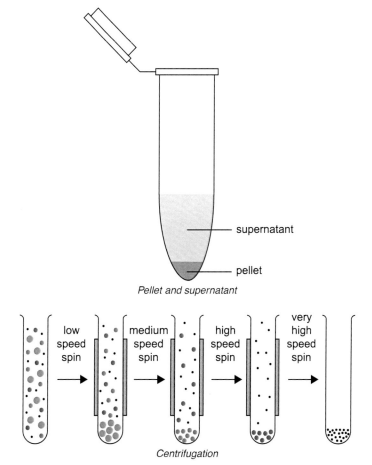

Pellet and supernatant

Centrifugation

A centrifuge can spin at extremely high speeds (up to 70,000 rpm for ultracentrifuges) which means that it must be carefully operated to avoid health and safety issues. For example, it is important to ensure that the tubes are balanced; this means that the samples placed into a centrifuge should hold a similar volume and should be placed in the rotor at opposite sides.

Gel electrophoresis

Proteins and nucleic acids can be separated by gel electrophoresis. This process separates macromolecules based upon their charge and/or size/shape. Gel electrophoresis uses current flowing through a buffer to separate macromolecules. The gel used in electrophoresis acts as a sieve, separating the macromolecules. One form of gel electrophoresis is SDS-PAGE. During this procedure, the proteins are denatured and given a uniform negative charge; this means that the proteins can be separated based on their size as they migrate towards the positive electrode. Small proteins travel further through the gel than large proteins.

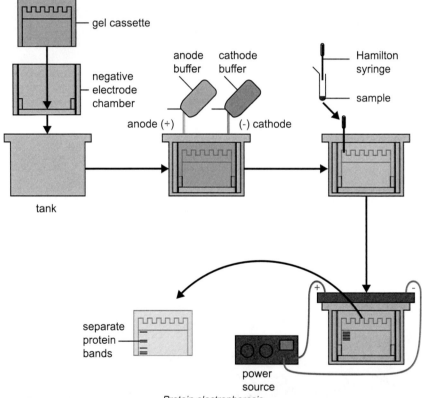

Protein electrophoresis

It is also possible to run native gels where the protein is not denatured before the gel electrophoresis. This allows the scientist to analyse the proteins in their folded state. In this case, the migration of the proteins through the gel depends on the protein shape, size and charge.

Iso-electric point

Proteins can be separated from a mixture using their isoelectric points (IEPs). IEP is the pH at which a soluble protein has no net charge and will precipitate out of solution. If the solution is buffered to a specific pH, only the protein(s) that have an IEP of that pH will precipitate.

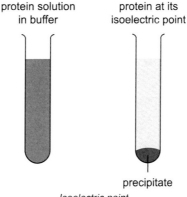

Isoelectric point

Proteins are made up of amino acids which may carry a positive, negative or neutral charge. The combined charges of the amino acids in a protein give the protein its overall charge. The charge of a protein will vary depending on the pH of the solution it is in. At their iso-electric point, proteins have an overall neutral charge. Proteins carry an overall positive charge at a pH below their isoelectric point and a negative charge at a pH above their isoelectric point. Proteins can also be separated using their IEPs in electrophoresis. Soluble proteins can be separated using an electric field and a pH gradient. A protein stops migrating through the gel at its IEP in the pH gradient because it has no net charge. This is known as isoelectric focussing.

In isoelectric focussing, a pH gradient is set up along a tube of polyacrylamide gel using a mixture of special buffers. Each protein in the sample that is loaded onto the gel will move until it reaches the pH corresponding to its isoelectric point. At this pH, the protein will move no further and form a band which can be visualised after staining.

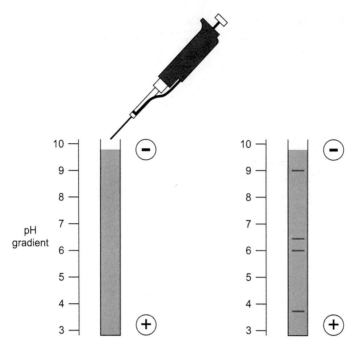

Isoelectric focussing

Paper, thin layer and affinity chromatography

Chromatography refers to a set of techniques which separates the components of a mixture. Chromatography allows scientists to identify and (in some types of chromatography) purify the components of a mixture. Chromatography can be used to separate mixtures of amino acids and sugars. The mixture is dissolved in a fluid known as the mobile phase. The components of the mixture are separated as they pass through a stationary phase, which varies depending on the type of chromatography being used.

In paper chromatography, the stationary phase is a strip of chromatography paper. A sample of the mixture being separated is placed in a dot, or line, near the bottom of a strip of chromatography paper which is then placed in a solvent. The solvent moves up through the chromatography paper and carries the components of the mixture with it, which will travel at different rates depending on their properties. For example, paper contains cellulose fibres which are polar in nature; any components of the mixture which are polar will bond with the cellulose fibres relatively quickly and do not travel far up the paper. Non-polar components of the mixture will not bind as readily to the paper and travel further.

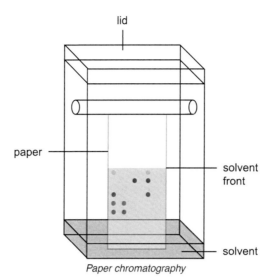
Paper chromatography

Thin layer chromatography (TLC) is similar to paper chromatography, but the stationary phase differs. Rather than using a stationary phase of chromatography paper, TLC uses a strip of absorbent material, such as silica gel, on a non-reactive backing. The rest of the process works in a similar manner to paper chromatography where a solvent moves up through the stationary phase and carries the components of the mixture with it. Again, the components will travel different distances depending on how soluble they are in the solvent and how much they bind to the stationary phase.

The process of affinity chromatography differs from paper and thin layer chromatography. Affinity chromatography relies on the binding interactions between a protein and specific molecules bound to a matrix or gel. A specific molecule is immobilised on an inert support in a column and a protein mixture is passed through the column. The target protein, which is complementary to the specific molecule in the column, will bind to it and remain in the column when the other components are washed away. The target protein can then be stripped from the support, resulting in its separation and purification from the original sample.

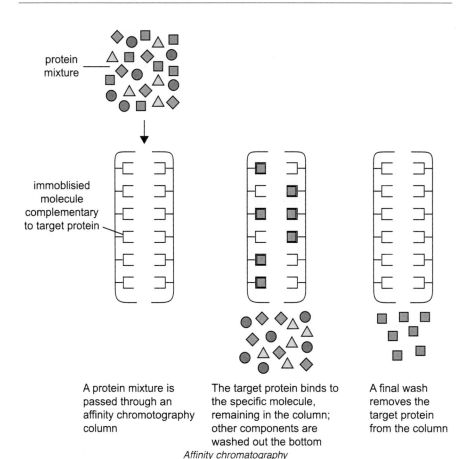

Affinity chromatography

1.4 Detecting proteins using antibodies

Immunoassay techniques

Antibodies play an important role within the immune system, identifying foreign proteins (antigens) and flagging them for destruction. Scientists have learned to produce antibodies and use them in the detection and identification of proteins; this group of procedures is known as immunoassay techniques.

Antibodies can be used to detect both the presence and concentration of a protein within a solution. **Immunoassays** rely on the specificity of antibodies; in other words, their ability to recognise and bind with only one molecule; these antibodies are known as monoclonal antibodies. Any antibody used in an immunoassay must be linked to a detectable label to allow scientists to detect when binding has occurred. These labels may be in the form of a reporter enzyme which causes a colour change in the presence of a specific antigen. Chemiluminescence (the production of light from a chemical reaction), fluorescence and other reporters can also be used.

The following diagram shows a typical immunoassay: the molecule being detected (a protein) is to the lower left, the antibody in the middle and the detectable label at the top.

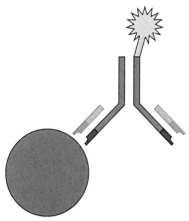

Immunoassay

In some cases, the assay uses a specific antigen to detect the presence of antibodies; for example in the detection of HIV. The immune system of a patient with HIV will produce antibodies which are complementary to HIV viral proteins; therefore, detecting the presence of these antibodies indicates a positive HIV result. To conduct this immunoassay, HIV proteins are mounted on a solid surface and a patient sample is added. If the patient has HIV, the antibodies produced by their immune system will bind to the HIV proteins on the solid surface. A secondary antibody, which is linked to a detectable label and is complementary to the patient's own antibodies, is added and indicates if the patient has HIV.

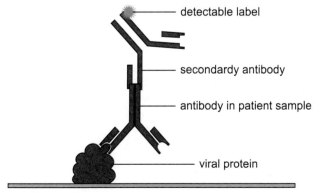

Detection of an antibody using immunoassay techniques

Western blotting

Western blotting is a technique used after SDS-PAGE electrophoresis. The separated proteins from the gel are transferred (blotted) onto a solid medium, often a nitrocellulose membrane.

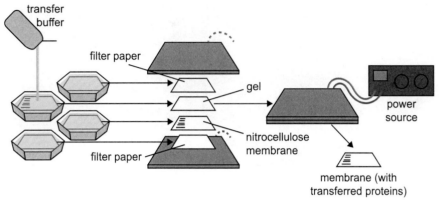

Protein blotting process

The membrane is then probed for the protein of interest using a specific antibody that is linked to a detectable label. This label may be a reporter enzyme which brings about a detectable colour change to indicate the presence of the target protein. Western blotting allows scientists to identify specific proteins which are present in a cell sample, for example if it has medical applications in the diagnosis of HIV and hepatitis B.

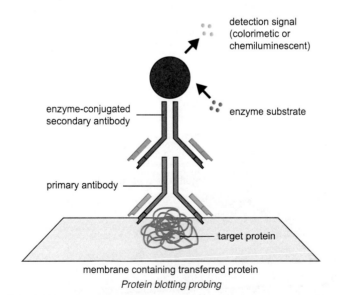

Protein blotting probing

1.5 Microscopy

Bright field and fluorescence microscopy

Bright-field microscopy is a relatively simple and straightforward microscopy technique. A sample is mounted on a slide and illuminated from below. Light is transmitted through the specimen to the objective lens (which magnifies the image) and then to the eyepiece at the top of the microscope where the image is observed. The image of the sample that is produced is usually darker than the background which appears bright, hence the name bright field microscopy. Samples are often stained before being viewed using a bright field microscope to increase contrast.

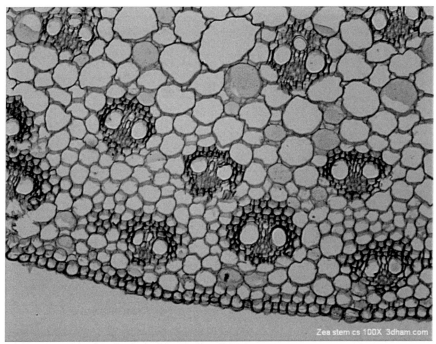

Bright field micrograph image of a stem cross section by John Alan Elson, licenced under the Creative Commons http://creativecommons.org/licenses/by-sa/3.0/deed.en license

Fluorescence microscopy allows particular protein structures to be visualised. A fluorescent molecule is one which absorbs a specific wavelength of light then emits a different (longer) wavelength. This means that it absorbs light of one colour and emits light of a different colour. In fluorescence microscopy, specific protein structures have fluorescent markers added to them. The cells can then be placed on a slide and the protein structure visualised using a fluorescence microscope.

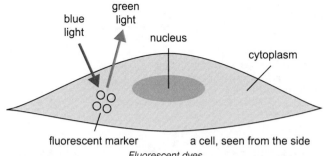

Fluorescent dyes

In some cases, antibodies are used to fluorescently tag protein structures. This is known as immunofluoresecence. A primary antibody, which is specific to the protein being visualised, is introduced to a cell sample. A secondary antibody, attached to a fluorescent tag, is then added which binds to the primary antibody.

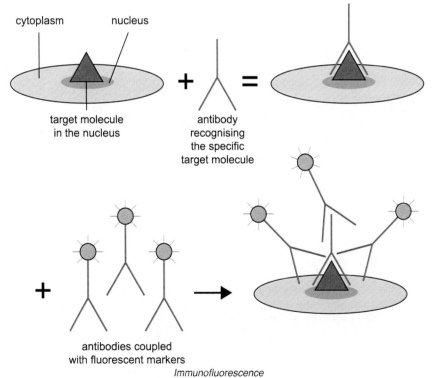

Immunofluorescence

In a fluorescent microscope, the light passing through the sample first passes through a filter which only lets through light at a specific wavelength that corresponds to the fluorescent marker being used. When this light hits the fluorescent marker, it fluoresces and emits light of a different wavelength. The second filter separates emitted light from the light first passed through the specimen and the fluorescing regions of the sample can be viewed in the microscope.

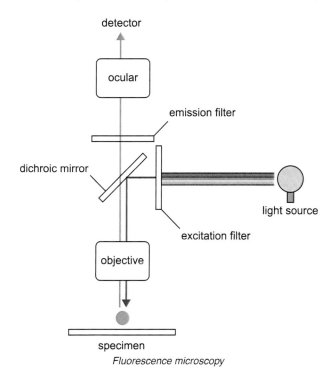

Fluorescence microscopy

The following image shows endothelial cells under a microscope. Nuclei are the small dark ovals, microtubules immediately surround them and actin filaments are found towards the edges.

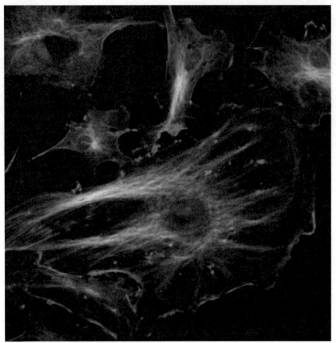
Fluorescent cells

1.6 Aseptic technique and cell culture

Cell culture

Cell culture is the ability to grow cells in an artificial laboratory environment. Cell culture is necessary for growing bacterial cells, culturing mammalian cells for cancer studies and many other processes.

Cell culture is often performed in flasks such as those shown in the following image. Cell culture requires environmental factors, for example temperature and pH, to be controlled and the growing cells must be given an opportunity for gas exchange. Many culture media exist that promote the growth of specific types of cells and microbes. Both plant and animal cell cultures require complex culture media which contain all the requirements of the cells. Typical culture media contains water, salts (Murashige and Skoog salts for plants), amino acids, vitamins and glucose. Animal cell culture also requires media containing growth factors from serum. Growth factors are proteins that promote cell growth and proliferation. Growth factors are essential for the culture of most animal cells. A microbial culture can be started using an inoculum of microbial cells on an agar medium, or in a broth with suitable nutrients.

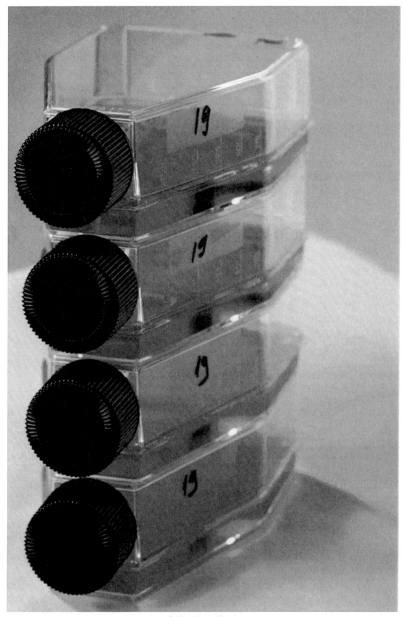
Cell culture flask

Most cells will only divide a limited number of times and then die. This makes cell culture a challenging procedure for primary cell lines. Tumour cell lines, on the other hand, can perform unlimited divisions and will grow and divide indefinitely in cell culture.

Aseptic technique

Aseptic techniques are a vital part of successful cell culture. Aseptic technique eliminates unwanted microbial contaminants when culturing micro-organisms or cells. Within all laboratory environments, there are many potential sources of contamination, for example non-sterile media and implements, airborne microorganisms and unclean work surfaces. There are many essential aspects to aseptic techniques which are detailed as follows.

Sterile work area

It is important to ensure that cell culture is performed in a sterile area. To achieve this, cell culture is often performed in culture hoods as shown in the following image. Surfaces are also disinfected before and after use with 70% ethanol or Virkon (a disinfectant).

Cell culture hood

Good personal hygiene

When performing cell culture, hands should be washed before and after all procedures. Appropriate personal protective equipment (PPE) should also be worn, for example a lab coat. This helps to avoid contaminants from the skin and clothes contaminating the cell culture. Gloves should also be worn and any cuts should be covered with a plaster.

Sterile reagents and media

It is important to ensure that all culture media and reagents are free from contaminants. Laboratories use commercial reagents and media which are sterilised as part of their production. They may need to be sterilised again once they have been handled in the laboratory, for example by using an autoclave which is a piece of equipment used to sterilise equipment and other supplies. The object to be sterilised is placed in the chamber of the autoclave and exposed to high pressure and temperature.

Sterile handling

As well as wiping surfaces with ethanol, containers, dishes and flasks should also be wiped down. All reagents should be exposed to the air for as short a period of time as possible, for example by replacing bottle lids swiftly. All pipettes which are used to transfer liquids are either sterile glass or single use disposable plastic. Experiments should be performed as quickly as possible to avoid the potential for entry of contaminants.

Counting microbial cells in culture

It is often useful to determine the number of cells in a liquid microbial culture. This can be achieved by plating out a liquid microbial culture on to a solid medium (such as nutrient agar) to allow the number of colony-forming units to be counted and the density of cells in the culture estimated. Serial dilution is often needed to achieve a suitable colony count as shown in the following diagram. Each colony arises from a single microbial cell; knowledge of the dilution factor then allows the number of cells in the original culture to be determined.

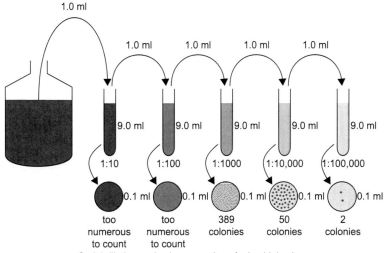

Serial dilution and colony counting of microbial culture

Haemocytometer

A **haemocytometer** is a specialised slide that has a counting chamber with a known volume of liquid; it allows an estimation of the concentration of cells in a culture to be made. A haemocytometer resembles a microscope slide, but has a grid made up of perpendicular lines (similar to graph paper) etched into the glass.

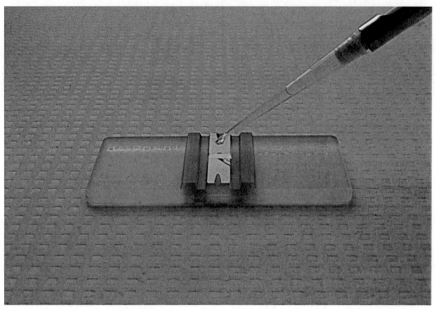

Haemocytometer by http://flickr.com/photos/83788754@N00, licenced under the Creative Commons http://creativecommons.org/licenses/by-sa/2.0/deed.en license

TOPIC 1. LABORATORY TECHNIQUES FOR BIOLOGISTS

To use the haemocytometer, a cover slide is affixed to it; this creates a chamber with a known depth (for example 0.1 mm). The cell culture to be counted is gently mixed. Some of this mixture is pipetted under the coverslip and the haemocytometer is placed under a microscope to visualise the grid. The following diagram shows the grid on a haemocytometer.

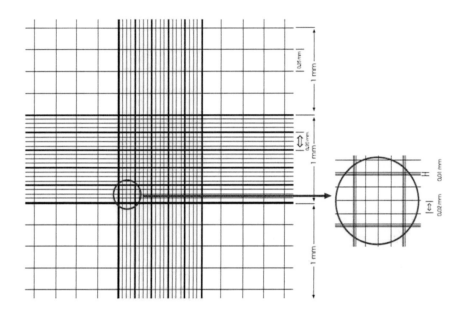

Haemocytometer grid by SantiBadia, licenced under the Creative Commons http://creativecommons.org/licenses/by-sa/2.0/deed.en license

The grid has nine 'large' squares, each 1 mm × 1 mm. Using this and the depth under the coverslip (0.1 mm), we can calculate the volume of cell culture in each area of the grid:

$$1 \times 1 \times 0.1 = 0.1 \text{ mm}^3 \text{ (or } 0.1 \text{ } \mu l\text{)}$$

The cells within one of the 1 mm × 1 mm areas are visualised using a microscope and counted. This provides an estimate of the number of cells per 0.1 μl of culture medium. Multiplying by 10 gives the number of cells per μl and multiplying by 10,000 gives the number of cells per cm^3.

The following diagram shows one of the 1 mm × 1 mm squares of a haemocytometer containing a cell culture sample.

UNIT 1. CELLS AND PROTEINS

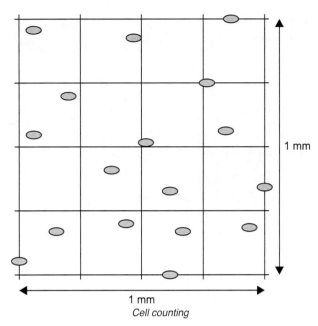

Cell counting

From the diagram, it is clear that there may be sources of error when using a haemocytometer to count cells. Should the cells touching the outer boundaries of the area be counted in or not? In general, a rule is set up where, if they are along the top or right border they are counted, whereas if they are along the left or bottom border they are not counted. Using the example above, we can estimate the number of cells per cm^3 of culture:

15 cells counted = 15 cells per 0.1 μl

$15 \times 10 = 150$ cells per μl

$150 \times 1000 = 150000 \; \left(1 \cdot 5 \times 10^5\right)$ cells per cm^3

To improve the reliability of the results, the number of cells in more than one square should be counted and an average calculated.

A haemocytometer can be used to provide an estimate of both the total and the viable cell number. A viable cell count identifies the number of actively growing/dividing cells in a sample. The haemocytometer is set up using cells stained with trypan blue dye, which is taken up by dead cells but not by living cells. A live cell count can then be performed where only unstained cells are counted. A total cell count is performed by counting all of the cells, including those stained with trypan blue. The percentage viability of a sample can be calculated using the following formula:

$$\frac{\text{Live cell count}}{\text{Total cell count}} = \text{Percentage viability}$$

e.g. $\dfrac{45 \times 10^4 \text{ ml}}{46 \times 10^4 \text{ ml}} = 97.8\%$ viability

1.7 Learning points

Summary

- Substances, organisms, and equipment in a laboratory can present a hazard.
- Hazards in the lab include toxic or corrosive chemicals, heat or flammable substances, pathogenic organisms and mechanical equipment.
- Risk is the likelihood of harm arising from exposure to a hazard.
- Risks can be controlled in the lab by carrying out a risk assessment to identify hazards and putting appropriate control measures in place to minimise the risk.
- Control measures include using appropriate handling techniques, protective clothing and equipment, and aseptic technique.
- A dilution series is a stepwise dilution of a stock solution.
- Dilutions in a linear dilution series differ by an equal interval, for example 0·1, 0·2, 0·3 and so on.
- Dilutions in a log dilution series differ by a constant proportion, for example 10^{-1}, 10^{-2}, 10^{-3} and so on.
- To determine the concentration of an unknown solution, a standard curve can be used. Plotting measured values for known concentrations to produce a line or curve allows the concentration of an unknown to be determined from the standard curve.
- Addition of acid or alkali has very small effects on the pH of a buffer, allowing the pH of a reaction mixture to be kept constant.
- A colorimeter measures the absorbance of specific wavelengths of light by a solution.
- To use a colorimeter, it is calibrated with an appropriate blank as a baseline. Absorbance measurements can then be used to determine concentration of a coloured solution using suitable wavelength filters or percentage transmission can be used to determine turbidity, such as cells in suspension.
- A centrifuge spins a sample at high speed to separate substances of differing density; more dense components settle in the pellet, less dense components remain in the supernatant.
- Paper and thin layer chromatography can be used for separating different substances such as amino acids and sugars.
- The speed that each solute travels along the chromatogram depends on its differing solubility in the solvent used.
- In affinity chromatography a solid matrix or gel column is created with specific molecules bound to the matrix or gel. Soluble, target proteins in a mixture, with a high affinity for these molecules, become attached to them as the mixture passes down the column. Other non-target molecules with a weaker affinity are washed out.
- Gel electrophoresis can be used to separate proteins and nucleic acids.

> **Summary continued**
> - During the gel electrophoresis process, charged macromolecules move though an electric field applied to a gel matrix.
> - Native gels do not denature the molecule so that separation is by shape, size and charge. Native gels separate proteins by their shape, size and charge.
> - SDS-PAGE gives all the molecules an equally negative charge and denatures them, separating proteins by size alone.
> - Proteins can be separated from a mixture using their isoelectric points (IEPs).
> - IEP is the pH at which a soluble protein has no net charge and will precipitate out of solution; therefore, if the solution is buffered to a specific pH, only the protein(s) that have an IEP of that pH will precipitate.
> - Proteins can also be separated using their IEPs in electrophoresis. Soluble proteins can be separated using an electric field and a pH gradient. A protein stops migrating through the gel at its IEP in the pH gradient because it has no net charge.
> - Immunoassay techniques are used to detect and identify specific proteins; these techniques use stocks of antibodies with the same specificity, known as monoclonal antibodies.
> - An antibody specific to the protein antigen is linked to a chemical 'label' to allow detection.
> - The 'label' is often a reporter enzyme producing a colour change, but chemiluminescence, fluorescence and other reporters can be used.
> - In some cases the assay uses a specific antigen to detect the presence of antibodies.
> - Western blotting is a technique, used after SDS-PAGE electrophoresis, the separated proteins from the gel are transferred (blotted) onto a solid medium and the proteins can be identified using specific antibodies that have reporter enzymes attached.
> - Bright-field microscopy is commonly used to observe whole organisms, parts of organisms, thin sections of dissected tissue or individual cells.
> - Fluorescence microscopy uses specific fluorescent labels to bind to and visualise certain molecules or structures within cells or tissues.
> - Aseptic technique eliminates unwanted microbial contaminants when culturing micro-organisms or cells.
> - Aseptic technique involves the sterilisation of equipment and culture media by heat or chemical means and subsequent exclusion of microbial contaminants.
> - A microbial culture can be started using an inoculum of microbial cells on an agar medium or in a broth with suitable nutrients.
> - Many culture media exist that promote the growth of specific types of cells and microbes.
> - Animal cells are grown in medium containing growth factors from serum.

> **Summary continued**
>
> - Growth factors are proteins that promote cell growth and proliferation. Growth factors are essential for the culture of most animal cells.
>
> - In culture, primary cell lines can divide a limited number of times, whereas tumour cells lines can perform unlimited divisions.
>
> - Plating out of a liquid microbial culture on solid media allows the number of colony-forming units to be counted and the density of cells in the culture estimated; serial dilution is often needed to achieve a suitable colony count.
>
> - A haemocytometer is a specialised slide that has a counting chamber with a known volume of liquid; it allows an estimation of the concentration of cells in a liquid culture to be made.
>
> - A viable cell count identifies the number of actively growing/dividing cells in a sample. The haemocytometer is set up using cells stained with trypan blue dye, which is taken up by dead cells but not by living cells - viable and total cell counts can then be performed and percentage viability calculated.

1.8 End of topic test

End of Topic 1 test Go online

A scientist working in a laboratory was carrying out a DNA gel electrophoresis and used ethidium bromide as a fluorescent tag to visualise the separated DNA. In order to make the ethidium bromide fluoresce and highlight the DNA bands, the gel was placed under UV light.

Read the following hazard information about ethidium bromide.

Ethidium bromide is strongly mutagenic. Ethidium bromide must also be considered a possible carcinogen and reproductive toxin. Therefore all individuals should regularly review their risk assessments and work practices for ethidium bromide. Ethidium bromide is readily absorbed through the skin. Ethidium bromide is highly toxic by inhalation, particularly in powder form, and is irritating to the skin, eyes, mucous membranes and upper respiratory tract.

Q1: Based on the information above, suggest two precautions which should be put in place when working with ethidium bromide. *(2 marks)*

....

Q2: Working with UV light in a lab may lead to certain risks for the user. Define the term risk. *(1 mark)*

....

Q3: Give an example of personal protective equipment (PPE) which would be appropriate for working with UV light. *(1 mark)*

A scientist performing a protein assay wanted to determine the concentration of protein in her sample. She performed a Bradford protein assay. Bradford reagent changes colour (and therefore absorbance) depending on the concentration of protein in the sample.

A buffer and Bradford reagent were added to the protein sample and the absorbance at 595 nm was measured.

Q4: Why are buffers used in experiments such as this? *(1 mark)*

....

Q5: Name a piece of equipment that is used to measure the absorbance of a solution. *(1 mark)*

....

Q6: Use the following standard curve to estimate the protein concentration of a sample with an absorbance of 0.6.

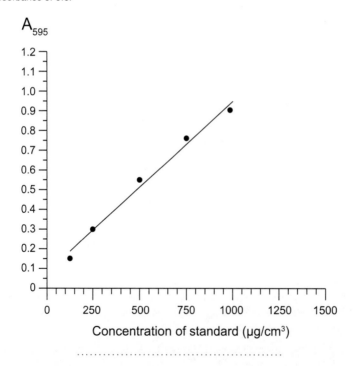

Q7: Name the group of techniques which are used to detect and identify specific proteins using antibodies. *(1 mark)*

....

Q8: Bright field microscopy can be used to view whole unicellular organisms. State another type of biological material that can be viewed using bright field microscopy. *(1 mark)*

Many biological research laboratories use cells to research biological processes and test new medicines. Cell culture is a complex process requiring many careful steps to be performed. When performing cell culture, cells are supplied with culture media which contains all the requirements of the cells.

Q9: What substance, containing growth factors, must be added to culture media to allow successful growth of animal cells in culture? *(1 mark)*

....

Q10: Why is it important to follow aseptic techniques when carrying out cell culture? *(1 mark)*

....

Q11: Other than wearing gloves and a lab coat, name one other aseptic technique. *(1 mark)*

The following diagram represents red blood cells in a haemocytometer viewed using a microscope. The grid is 0·1 mm in depth.

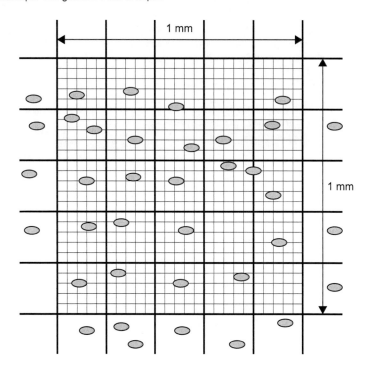

Q12: Calculate the number of red blood cells per cm^3 of blood sample. *(1 mark)*

....

Q13: What type of staining is required to identify and count viable cells? *(1 mark)*

Unit 1 Topic 2

Proteomics, protein structure, binding and conformational change

Contents

- 2.1 The proteome 41
- 2.2 The synthesis and transport of proteins 42
- 2.3 Protein structure 49
- 2.4 Ligand binding 57
- 2.5 Reversible binding of phosphate and the control of conformation 59
- 2.6 Learning points 61
- 2.7 Extended response questions 63
- 2.8 End of topic test 63

Prerequisites

You should already know that:

- proteins are held in a three-dimensional shape;
- amino acids are linked by peptide bonds to form polypeptides;
- polypeptide chains fold to form the three-dimensional shape of a protein, which is held together by hydrogen bonds and other interactions between individual amino acids;
- the role of the active site in orientating reactants is lowering the activation energy of the transition state and the release of products with low affinity for the active site;
- there are three types of inhibition: competitive inhibition (binds to active site), non-competitive inhibition (changes shape of active site) and feedback inhibition (end product binds to an enzyme which catalyses a reaction earlier in the pathway).

Learning objective

By the end of this topic, you should be able to:

- state that the proteome is the entire set of proteins expressed by a genome;
- explain why the proteome is larger than the number of genes in an organism;
- state that not all genes are expressed as proteins in a particular cell type and give examples;
- state that the set of proteins expressed by a given cell type can vary over time and under different conditions;
- give examples of factors which affect the set of proteins expressed by a given cell;
- state that eukaryotic cells have a system of internal membranes, which increases the total area of membrane;
- describe the structure of the endoplasmic reticulum (ER), Golgi apparatus and lysosomes;
- describe the role of vesicles;
- state that lipids and proteins are synthesised in the ER;
- describe the difference between the rough ER (RER) and smooth ER (SER);
- give the location where lipids are produced;
- give the location where cytosolic proteins are produced;
- describe the location of transmembrane protein production;
- state that once the proteins are in the ER, they are transported by vesicles that bud off from the ER and fuse with the Golgi apparatus;
- state that as proteins move through the Golgi apparatus they undergo post-translational modification and the addition of carbohydrate groups is the major modification;
- state that vesicles that leave the Golgi apparatus take proteins to the plasma membrane and lysosomes;
- state that vesicles move along microtubules to other membranes and fuse with them within the cell;
- give the location of translation of proteins which are secreted from the cell;
- describe the secretory pathway;
- state that many secreted proteins are synthesised as inactive precursors and require proteolytic cleavage to produce active proteins;
- state that digestive enzymes are one example of secreted proteins that require proteolytic cleavage to become active;

TOPIC 2. PROTEOMICS, PROTEIN STRUCTURE, BINDING AND CONFORMATIONAL CHANGE

Learning objective continued

- state that amino acid sequence determines protein structure and proteins are polymers of amino acid monomers;
- name the bonds which link amino acids together in proteins and recognise its chemical structure;
- describe the basic structure of an amino acid;
- classify amino acids according to the R group present;
- state that the wide range of functions carried out by proteins results from the diversity of R groups;
- describe the primary sequence of a protein;
- describe the secondary sequence of a protein, including alpha helices, parallel or anti-parallel beta-pleated sheets, or turns;
- state that the polypeptide folds into a tertiary structure;
- describe the interactions between R groups which stabilise tertiary structure;
- state that quaternary structure exists in proteins with two or more connected polypeptide subunits;
- state that a prosthetic group is a non-protein unit tightly bound to a protein and necessary for its function, and use the haem group of haemoglobin as an example;
- describe how Interactions of the R groups can be influenced by temperature and pH;
- define the term ligand;
- state that R groups not involved in protein folding can allow binding to ligands;
- state that:
  - binding sites will have complementary shape and chemistry to the ligand;
  - as a ligand binds to a protein-binding site, the conformation of the protein changes;
  - this change in conformation causes a functional change in the protein.
- describe how allosteric interactions occur between spatially distinct sites;
- state that many allosteric proteins consist of multiple subunits (quaternary structure);
- describe the process of co-operativity in binding;
- state that allosteric enzymes contain a second type of site, called an allosteric site, and that modulators regulate the activity of the enzyme when they bind to the this site;
- describe the effect the binding of a modulator may have on a protein;
- describe co-operativity in terms of the binding and release of oxygen from haemoglobin;
- describe the influence and physiological importance of temperature and pH on the binding of oxygen;

© HERIOT-WATT UNIVERSITY

Learning objective continued

- state that the addition or removal of phosphate can cause reversible conformational change in proteins - this is a common form of post-translational modification;
- describe the role of protein kinases;
- describe the use of ATP in phosphorylation reactions;
- describe the role of protein phosphatases;
- state that phosphorylation brings about conformational changes which can affect the activity of a protein and that the activity of many cellular proteins, such as enzymes and receptors, is regulated in this way;
- state that some proteins are activated by phosphorylation while others are inhibited;
- state that adding a phosphate group adds negative charges and that ionic interactions in the unphosphorylated protein can be disrupted and new ones created.

2.1 The proteome

The genome is all of the hereditary information encoded in DNA. The proteome is the entire set of proteins expressed by a genome. The proteome is larger than the number of genes, particularly in eukaryotes, because more than one protein can be produced from a single gene as a result of alternative RNA splicing. The set of proteins expressed by a given cell type can vary over time and under different conditions. Some factors affecting the set of proteins expressed by a given cell type are the metabolic activity of the cell, cellular stress, the response to signalling molecules, and diseased versus healthy cells.

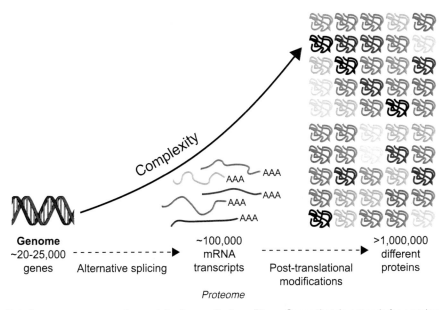

Genome ~20-25,000 genes — Alternative splicing → ~100,000 mRNA transcripts — Post-translational modifications → >1,000,000 different proteins

Proteome

Not all genes are expressed as proteins in a particular cell type. Genes that do not code for proteins are called non-coding RNA genes and include those that are transcribed to produce tRNA, rRNA, and RNA molecules that control the expression of other genes.

2.2 The synthesis and transport of proteins

Intracellular membranes

Eukaryotic cells have a relatively small surface area to volume ratio as a result of their size. The plasma membrane of eukaryotic cells is therefore too small an area to carry out all the vital functions carried out by membranes. Eukaryotic cells have a system of internal membranes, which increases the total area of membrane and provides a larger surface area for vital functions to take place.

The endoplasmic reticulum (ER) forms a network of membrane tubules continuous with the nuclear membrane. Rough ER (RER) has ribosomes on its cytosolic face while smooth ER (SER) lacks ribosomes.

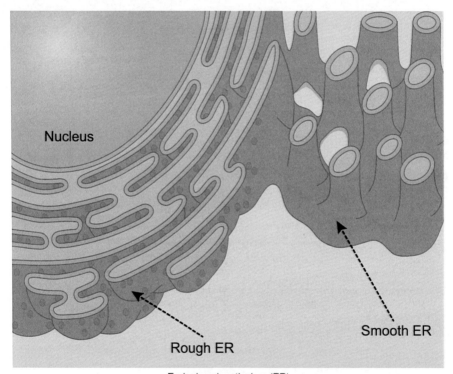

Endoplasmic reticulum (ER)

The Golgi apparatus is a series of flattened membrane discs. The discs are connected allowing molecules to move within the Golgi apparatus. The Golgi apparatus is adjacent to the endoplasmic reticulum.

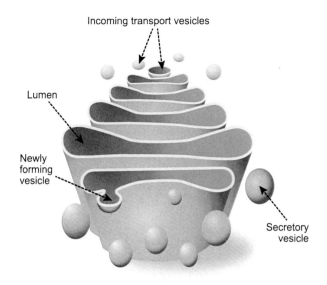

Golgi apparatus

Lysosomes are membrane-bound organelles containing a variety of hydrolases that digest proteins, lipids, nucleic acids and carbohydrates. The interior of the lysosomes are acidic allowing optimal function of the enzymes it contains.

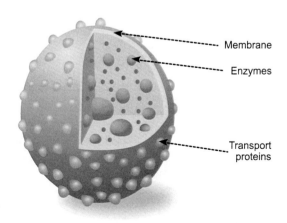

Lysosome

Vesicles transport materials between membrane compartments. They consist of an aqueous solution enclosed by a lipid bilayer.

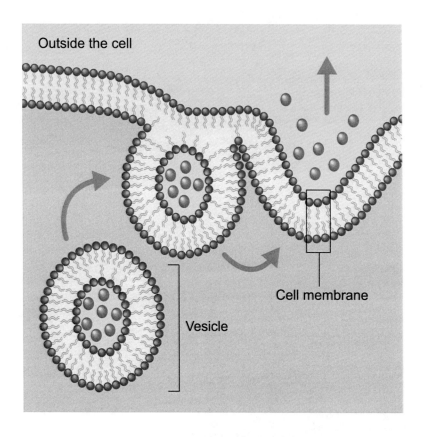

Vesicle

TOPIC 2. PROTEOMICS, PROTEIN STRUCTURE, BINDING AND CONFORMATIONAL CHANGE

The following diagram summarises the eukaryotic endomembrane system.

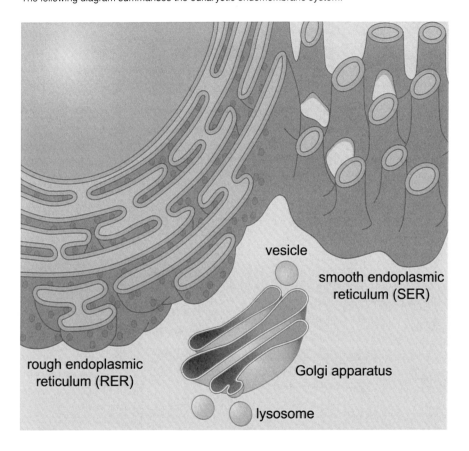

Intracellular membranes

Synthesis of membrane components

Lipids and proteins are synthesised in the ER. There are two types of ER within eukaryotic cells; rough ER (RER) has ribosomes on its cytosolic face while smooth ER (SER) lacks ribosomes.

Lipids are synthesised in the smooth endoplasmic reticulum (SER) and inserted into its membrane.

The synthesis of all proteins begins in cytosolic ribosomes. The synthesis of cytosolic proteins is completed there, and these proteins remain in the cytosol. Transmembrane proteins carry a signal sequence, which halts translation and directs the ribosome synthesising the protein to dock with the ER, forming RER. This signal sequence is a short stretch of amino acids at one end of the polypeptide that determines the eventual location of a protein in a cell. Translation continues after docking, and the protein is inserted into the membrane of the ER.

© HERIOT-WATT UNIVERSITY

46 UNIT 1. CELLS AND PROTEINS

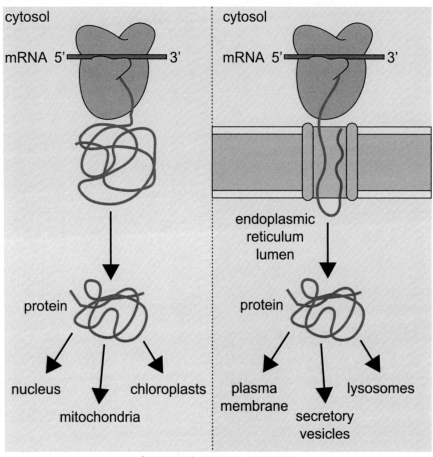

Synthesis of membrane components

Movement of proteins between membranes

Once the proteins are in the ER, they are transported by vesicles that bud off from the ER and fuse with the Golgi apparatus. As proteins move through the Golgi apparatus they undergo post-translational modification. Post-translational modification refers to covalent modifications which are made to proteins after translation. Molecules move through the Golgi discs in vesicles that bud off from one disc and fuse to the next one in the stack, as they move post-translational modifications take place. The addition of carbohydrate groups is the major post-translational modification; enzymes catalyse the addition of various sugars in multiple steps to form the carbohydrates. Various other chemical groups can also be added as summarised in the diagram.

TOPIC 2. PROTEOMICS, PROTEIN STRUCTURE, BINDING AND CONFORMATIONAL CHANGE

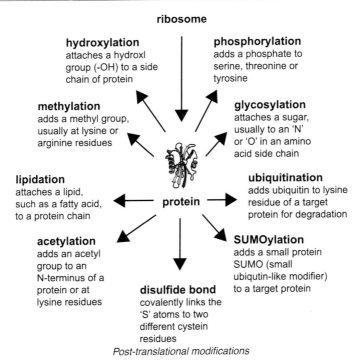

Post-translational modifications

Vesicles that leave the Golgi apparatus take proteins to the plasma membrane and lysosomes. Vesicles move along microtubules to other membranes and fuse with them within the cell. The following diagram illustrates that the vesicle is attached to a motor protein and moves along the microtubule of the cytoskeleton using energy from ATP hydrolysis.

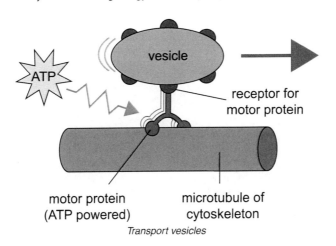

Transport vesicles

The secretory pathway

Secreted proteins are translated in ribosomes on the RER and enter its lumen. Peptide hormones (such as insulin) and digestive enzymes (such as trypsin) are examples of secreted proteins. The proteins move through the Golgi apparatus and are then packaged into secretory vesicles. These vesicles move to and fuse with the plasma membrane, releasing the proteins out of the cell.

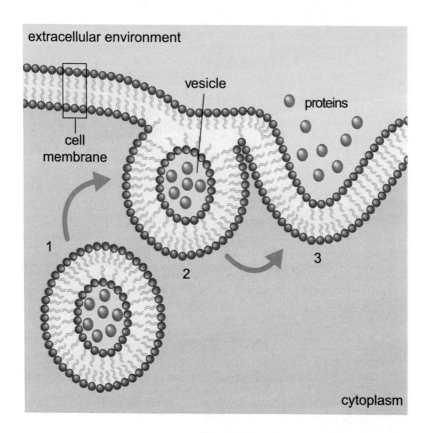

Vesicles

Many secreted proteins are synthesised as inactive precursors and require proteolytic cleavage to produce active proteins. Proteolytic cleavage is another type of post-translational modification. Digestive enzymes are one example of secreted proteins that require proteolytic cleavage to become active. For example the digestive enzyme chymotrypsin is produced in its inactive form, chymotrypsinogen, which must be proteolytically cleaved to become active chymotrypsin.

2.3 Protein structure

Amino acids

Amino acids are the building blocks of proteins; amino acid sequence determines protein structure. A monomer is a molecule that may bind chemically to other molecules to form a **polymer**; proteins are polymers of amino acid **monomers**. The amino acids in a protein are linked by peptide bonds to form **polypeptides**. Peptide bonds are covalent and very strong, they are difficult to break.

Peptide bond

Amino acids have the same basic structure, differing only in the R group present. Amino acids can be grouped according to their properties. All amino acids have a central carbon with four groups attached (an amine (NH_2), a carboxylic acid (COOH), a hydrogen and a variable R group).

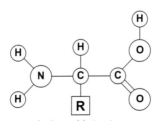

Amino acid structure

Amino acids are classified according to their R groups:

- basic (positively charged);
- acidic (negatively charged);
- polar;
- hydrophobic.

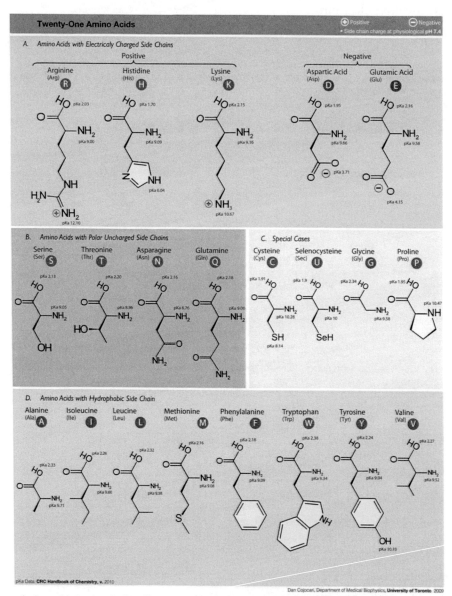

Amino acid structure by http://commons.wikimedia.org/wiki/User:Dancojocari, licenced under the Creative Commons http://creativecommons.org/licenses/by-sa/3.0/deed.en license

TOPIC 2. PROTEOMICS, PROTEIN STRUCTURE, BINDING AND CONFORMATIONAL CHANGE

Acidic (negatively charged) amino acids are hydrophilic and the key component of their R group is a carboxylic acid group (negatively charged at pH 7). Basic (positively charged) amino acids are hydrophilic and the key component of their R group is an amine group (positively charged at pH 7). Polar amino acids are hydrophilic and the key component of their R group are hydrophilic groups, like carbonyl (C=O), hydroxyl (OH) or amine (NH) groups. Hydrophobic amino acids are non-polar and the key component of their R group is a hydrocarbon group.

R groups of amino acids vary in size, shape, charge, hydrogen bonding capacity and chemical reactivity. The wide range of functions carried out by proteins results from the diversity of R groups.

Protein structure

The primary structure is the sequence in which the amino acids are synthesised into the polypeptide.

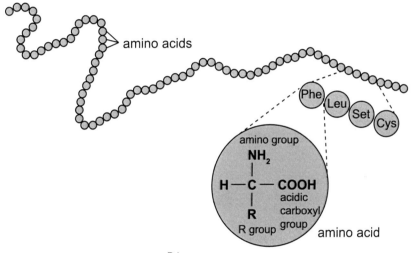

Primary sequence

The amino acids along the length of the polypeptide chain interact with one another. Some amino acids form hydrogen bonds which result in secondary protein structure. The secondary structure of a protein is stabilised by hydrogen bonds between atoms of the same chain. α-helix is one type of secondary structure. It is a spiral with the R groups sticking outwards. Another type of secondary structure is β-sheet. β-sheet has parts of the chain running alongside each other, forming a sheet. The R groups sit above and below the sheet.

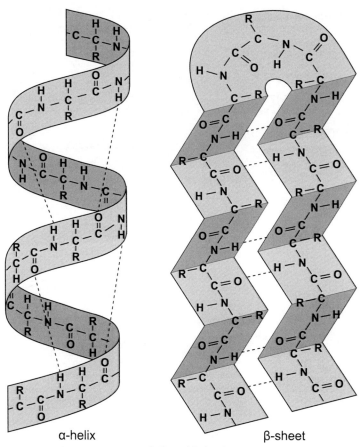

α-helix and β-sheet

β-sheets are usually anti-parallel, meaning that the chains run in opposite directions from each other. They can also be parallel, meaning that the chains run in the same direction. The sheets are parallel or antiparallel depending on their N and C termini as shown in the following diagram.

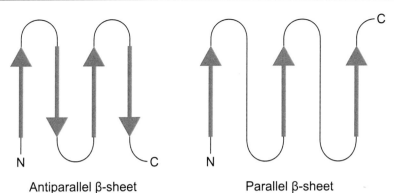

Antiparallel β-sheet   Parallel β-sheet

Antiparallel and parallel β-sheet

Turns are a third type of secondary structure; they reverse the direction of the polypeptide chain. The exact role of turns has not yet been determined, but some scientists believe that turns allow interactions between regular secondary structure elements.

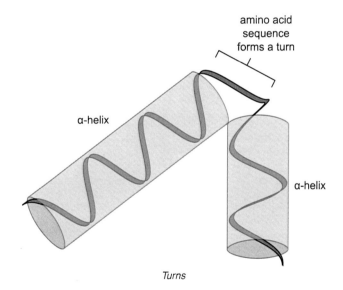

Turns

The polypeptide folds into a tertiary structure. Folding at this level is stabilised by many different interactions between the R groups of the amino acids.

Tertiary structure is brought about by charge effects, such as interactions of the R groups in hydrophobic regions. Hydrophobic amino acids tend to cluster together on the interior of a protein, away from the surface. Hydrophilic amino acids will predominate at the surface of a soluble protein. This hydrophobic effect is one of the main driving forces of protein folding.

Ionic, hydrogen and disulfide bonds are also involved in tertiary structure. In an ionic bond, atoms are oppositely charged and, therefore, held by an electrostatic attraction. A hydrogen bond is an electrostatic attractive interaction which occurs between a hydrogen atom and an electronegative atom, such as oxygen or nitrogen. A disulfide bond (also known as a disulfide bridge) is a covalent bond between two thiol (SH) groups.

One final charge effect which influences the tertiary structure of a protein is London dispersion forces. The London dispersion force is a temporary attractive force that results when the electrons in two adjacent atoms occupy positions that make the atoms form temporary dipoles.

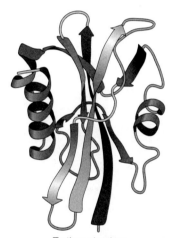

Tertiary structure

Many proteins are made up of more than one subunit. Quaternary structure exists in proteins with two or more connected polypeptide subunits. Quaternary structure describes the spatial arrangement of the subunits. For example, haemoglobin is made of four subunits as shown in the following diagram.

TOPIC 2. PROTEOMICS, PROTEIN STRUCTURE, BINDING AND CONFORMATIONAL CHANGE

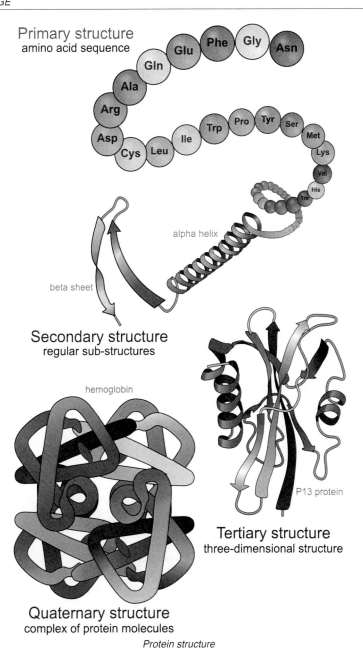

Protein structure

Prosthetic groups

Some proteins include prosthetic (non-protein) parts. A **prosthetic group** is a non-protein unit tightly bound to a protein and necessary for its function, e.g. haem in haemoglobin. This haem group contains an iron atom which is the site of oxygen binding. The iron atom is covalently bound to the haemoglobin via a series of histidine amino acids.

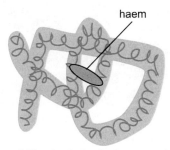

One haemoglobin subunit showing haem prosthetic group

Influence of temperature and pH on amino acid R groups

Interactions of the R groups can be influenced by pH and temperature. This is why pH and temperature will affect the structure (and function) of a protein. Increasing temperature disrupts the interactions that hold the protein in shape; the protein begins to unfold, eventually becoming denatured.

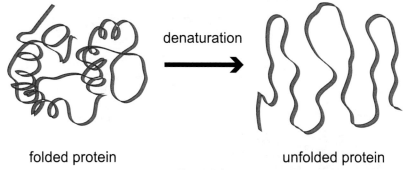

Denaturation

The charges on acidic and basic R groups are affected by pH. As pH increases or decreases from the optimum, the normal ionic interactions between charged groups are lost, which gradually changes the conformation of the protein until it becomes denatured.

2.4 Ligand binding

A **ligand** is a substance that can bind to a protein. R groups not involved in protein folding can allow binding to ligands. Binding sites will have complementary shape and chemistry to the ligand. As a ligand binds to a protein-binding site the conformation of the protein changes. This change in conformation causes a functional change in the protein.

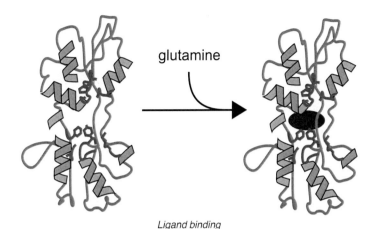

Ligand binding

Many allosteric proteins consist of multiple subunits (have quaternary structure). Allosteric interactions occur between spatially distinct sites; this means that the binding of a substrate molecule to one **active site** of an allosteric enzyme increases the affinity of the other active sites for binding of subsequent substrate molecules. This is of biological importance because the activity of allosteric enzymes can vary greatly with small changes in substrate concentration.

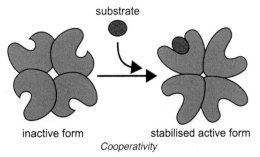

Cooperativity

Haemoglobin is a protein which shows cooperativity. Haemoglobin demonstrates quaternary structure in that is made up of four polypeptide subunits, each of which contain a haem group capable of binding a molecule of oxygen. When one of the subunits binds a molecule of oxygen, the second binds more easily, and the third and fourth easier still. This process is known as cooperativity; the ligand binding to one subunit of the protein has increased the other subunits' affinity for the ligand. When oxy-haemoglobin releases oxygen the same process happens; once one subunit has released its oxygen, the next subunit is more likely to release its oxygen and so on.

The main factors which will affect haemoglobin's ability to bind oxygen are:

- temperature - as temperature increases, affinity for oxygen decreases, curve shifts right;
- pH - as pH decreases, affinity for oxygen decreases, curve shifts right.

The following graph shows the partial pressure of oxygen in the blood versus oxyhaemglobin saturation (known as the oxygen-haemoglobin dissociation curve). The blue line represents standard physiological conditions, the red line represents the effect of decreasing pH or increasing temperature, and the green line represents the effect of increasing pH or decreasing temperature.

At a partial oxygen pressure (PO2) of 50 mmHg in normal physiological conditions, approximately 80% of haemoglobin is saturated; a decrease in pH or an increase in temperature reduces this value to nearer 75%. This change in saturation is caused by a decrease in haemoglobin's affinity for oxygen. A decrease in pH or an increase in temperature lowers the affinity of haemoglobin for oxygen, so the binding of oxygen is reduced. Reduced pH and increased temperature in actively respiring tissue will reduce the binding of oxygen to haemoglobin promoting increased oxygen delivery to tissue.

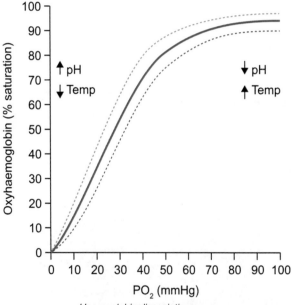

Haemoglobin dissociation curve

An allosteric enzyme is one which changes conformation upon binding a modulator. In allosteric enzymes, modulators bind at secondary binding sites, known as allosteric sites. These allosteric sites are separate and distinct from the active site of the enzyme. Modulators regulate the activity of the enzyme when they bind to the allosteric site. Upon binding a modulator, the conformation of an allosteric enzyme will change and this alters the affinity of the active site for the substrate. Modulators may be positive or negative. Negative modulators reduce the enzyme's affinity for the substrate and positive modulators increase the enzyme's affinity for the substrate.

TOPIC 2. PROTEOMICS, PROTEIN STRUCTURE, BINDING AND CONFORMATIONAL CHANGE

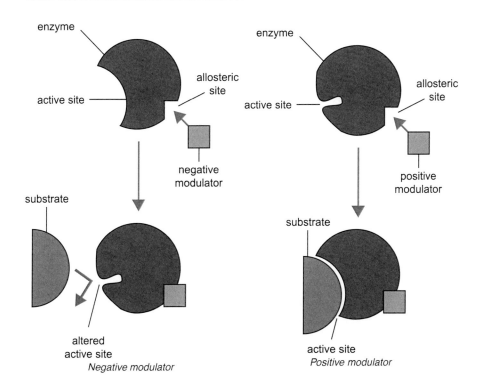

2.5 Reversible binding of phosphate and the control of conformation

Phosphorylation of proteins is a form of post-translational modification. The addition or removal of phosphate from particular R groups can be used to cause reversible conformational changes in proteins which can affect a protein's activity. Phosphorylation/dephosphorylation allows the activity of many cellular proteins, such as enzymes and receptors, to be regulated.

Protein kinases catalyse the transfer of a phosphate group to other proteins. As shown in the diagram below, the terminal phosphate of ATP is transferred to specific R groups creating ADP as well as a phosphorylated protein. Protein phosphatases catalyse the reverse reaction.

Some proteins are activated by phosphorylation while others are inhibited. For example phosphorylation of the enzyme glycogen phosphorylase b converts it to the active glycogen phosphorylase a; while phosphorylation of another metabolic enzyme called glycogen synthase inactivates the enzyme.

60 UNIT 1. CELLS AND PROTEINS

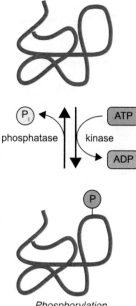

Phosphorylation

Adding a phosphate group adds negative charges in the side chain of amino acids. This can change a protein's structure by altering interactions with nearby amino acids; ionic interactions in the unphosphorylated protein can be disrupted and new ones created. Proteins can be phosphorylated and dephosphorylated with ease therefore this mechanism is often used to activate or deactivate proteins in response to changes in environmental conditions or external signals.

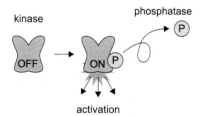

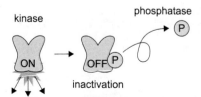

Protein activation/inactivation by phosphorylation

© HERIOT-WATT UNIVERSITY

2.6 Learning points

Summary

- Amino acid sequence determines protein structure.
- Proteins are polymers of amino acid monomers.
- Amino acids are linked by peptide bonds to form polypeptides.
- Amino acids have the same basic structure, differing only in the R group present.
- R groups of amino acids vary in size, shape, charge, hydrogen bonding capacity and chemical reactivity.
- Amino acids are classified according to their R groups: basic (positively charged); acidic (negatively charged); polar; hydrophobic.
- The wide range of functions carried out by proteins results from the diversity of R groups.
- The primary structure is the sequence in which the amino acids are synthesised into the polypeptide.
- Hydrogen bonding along the backbone of the protein strand results in regions of secondary structure - alpha helices, parallel or anti-parallel beta-pleated sheets, or turns.
- The polypeptide folds into a tertiary structure; this conformation is stabilised by interactions between R groups: hydrophobic interactions; ionic bonds; London dispersion forces; hydrogen bonds; disulfide bridges.
- Disulfide bridges are covalent bonds between R groups containing sulfur.
- Quaternary structure exists in proteins with two or more connected polypeptide subunits.
- Quaternary structure describes the spatial arrangement of the subunits.
- A prosthetic group is a non-protein unit tightly bound to a protein and necessary for its function.
- The ability of haemoglobin to bind oxygen is dependent upon the non-protein haem group.
- Interactions of the R groups can be influenced by temperature and pH.
- Increasing temperature disrupts the interactions that hold the protein in shape; the protein begins to unfold, eventually becoming denatured.
- The charges on acidic and basic R groups are affected by pH. As pH increases or decreases from the optimum, the normal ionic interactions between charged groups are lost, which gradually changes the conformation of the protein until it becomes denatured.
- A ligand is a substance that can bind to a protein.

Summary continued

- R groups not involved in protein folding can allow binding to ligands.
- Binding sites will have complementary shape and chemistry to the ligand.
- As a ligand binds to a protein-binding site the conformation of the protein changes.
- This change in conformation causes a functional change in the protein.
- Allosteric interactions occur between spatially distinct sites.
- The binding of a substrate molecule to one active site of an allosteric enzyme increases the affinity of the other active sites for binding of subsequent substrate molecules - this is of biological importance because the activity of allosteric enzymes can vary greatly with small changes in substrate concentration.
- Many allosteric proteins consist of multiple subunits (have quaternary structure).
- Allosteric proteins with multiple subunits show co-operativity in binding, in which changes in binding at one subunit alter the affinity of the remaining subunits.
- Allosteric enzymes contain a second type of site, called an allosteric site.
- Modulators regulate the activity of the enzyme when they bind to the allosteric site.
- Following binding of a modulator, the conformation of the enzyme changes and this alters the affinity of the active site for the substrate.
- Positive modulators increase the enzyme's affinity for the substrate, whereas negative modulators reduce the enzyme's affinity.
- The binding and release of oxygen in haemoglobin shows co-operativity.
- Changes in binding of oxygen at one subunit alter the affinity of the remaining subunits for oxygen.
- A decrease in pH or an increase in temperature lowers the affinity of haemoglobin for oxygen, so the binding of oxygen is reduced. Reduced pH and increased temperature in actively respiring tissue will reduce the binding of oxygen to haemoglobin promoting increased oxygen delivery to tissue.
- The addition or removal of phosphate can cause reversible conformational change in proteins; this is a common form of post-translational modification.
- Protein kinases catalyse the transfer of a phosphate group to other proteins.
- The terminal phosphate of ATP is transferred to specific R groups.
- Protein phosphatases catalyse the reverse reaction.
- Phosphorylation brings about conformational changes, which can affect a protein's activity.
- The activity of many cellular proteins, such as enzymes and receptors, is regulated in this way.

TOPIC 2. PROTEOMICS, PROTEIN STRUCTURE, BINDING AND CONFORMATIONAL CHANGE

> **Summary continued**
>
> - Some proteins are activated by phosphorylation while others are inhibited.
> - Adding a phosphate group adds negative charges. Ionic interactions in the unphosphorylated protein can be disrupted and new ones created.

2.7 Extended response questions

The activities which follow present extended response questions similar to the style that you will encounter in the examination.

You should have a good understanding of the structure of proteins and enzyme activation before attempting the questions.

You should give your completed answers to your teacher or tutor for marking, or try to mark them yourself using the suggested marking schemes.

Extended response question: Structure of proteins

Give details of the structure of proteins including primary, secondary, tertiary and quaternary levels. *(10 marks)*

Extended response question: Enzyme activation

Give an account of enzyme activation. *(5 marks)*

2.8 End of topic test

End of Topic 2 test Go online

Q1: What is a proteome? *(1 mark)*

....

Q2: What name is given to genes that do not code for proteins? *(1 mark)*

....

Q3: Identify one factor which can affect the set of proteins expressed by a given cell type. *(1 mark)*

The following illustration shows a magnified view of an animal cell.

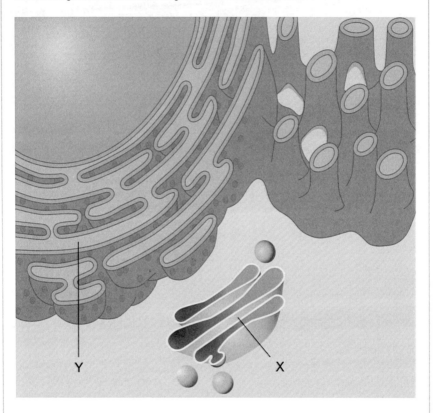

Q4: Identify organelle X. *(1 mark)*
....
Q5: Identify organelle Y. *(1 mark)*
....
Q6: Identify the location of lipid synthesis within the cell. *(1 mark)*
....
Q7: Describe how vesicles are able to move around the cell. *(1 mark)*
....
Q8: Give one example of a post-translational modification. *(1 mark)*

TOPIC 2. PROTEOMICS, PROTEIN STRUCTURE, BINDING AND CONFORMATIONAL CHANGE

Q9: What class does the following amino acid belong to? *(1 mark)*

$$\begin{array}{c} OH \\ | \\ H \quad CH_2 \quad OH \\ \backslash \quad | \quad / \\ N-C-C \\ / \quad | \quad \backslash \\ H \quad H \quad O \end{array}$$

....

Q10: What class does the following amino acid belong to? *(1 mark)*

$$\begin{array}{c} O \quad OH \\ \backslash\!\!/ \\ C \\ | \\ H \quad CH_2 \quad OH \\ \backslash \quad | \quad / \\ N-C-C \\ / \quad | \quad \backslash \\ H \quad H \quad O \end{array}$$

....

Q11: What is the primary protein sequence? *(1 mark)*

The following illustration shows the production of active insulin from proinsulin.

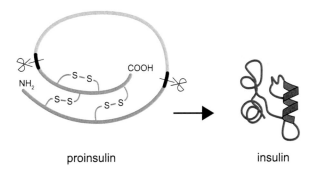

Q12: What name is given to the covalent bonds which link sulfur atoms as shown in the diagram? *(1 mark)*

....

Q13: A section of the proinsulin chain is cut and removed to form active insulin, what term described this process? *(1 mark)*

....

Q14: What aspect(s) of the secondary structure of proteins can be seen in the illustration? *(1 mark)*

....

Q15: What is a ligand? *(1 mark)*

....

Q16: Some enzymes have secondary binding sites which are spatially distinct from the active site. What name is given to these site? *(1 mark)*

The following diagram shows co-operativity in binding of oxygen by haemoglobin.

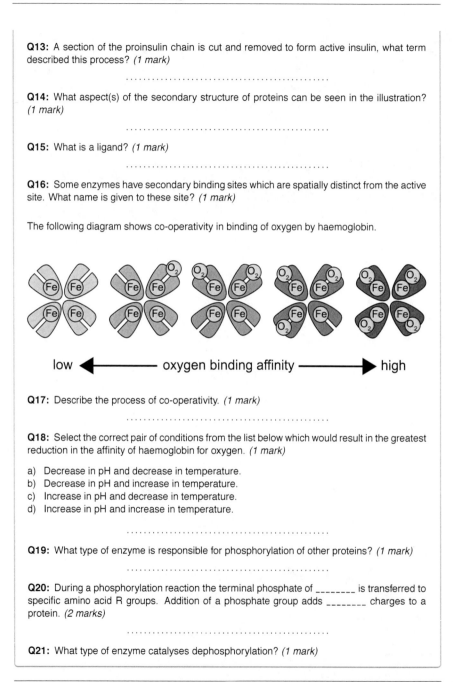

Q17: Describe the process of co-operativity. *(1 mark)*

....

Q18: Select the correct pair of conditions from the list below which would result in the greatest reduction in the affinity of haemoglobin for oxygen. *(1 mark)*

a) Decrease in pH and decrease in temperature.
b) Decrease in pH and increase in temperature.
c) Increase in pH and decrease in temperature.
d) Increase in pH and increase in temperature.

....

Q19: What type of enzyme is responsible for phosphorylation of other proteins? *(1 mark)*

....

Q20: During a phosphorylation reaction the terminal phosphate of _____ is transferred to specific amino acid R groups. Addition of a phosphate group adds _____ charges to a protein. *(2 marks)*

....

Q21: What type of enzyme catalyses dephosphorylation? *(1 mark)*

Unit 1 Topic 3

Membrane proteins

Contents

3.1 Movement of molecules across membranes 69
3.2 Ion transport pumps and generation of ion gradients 75
3.3 Learning points 77
3.4 Extended response question 79
3.5 End of topic test 80

Prerequisites

You should already know that:

- cells are specialised to perform specific functions;
- passive processes do not require energy;
- active processes require energy.

Learning objective

By the end of this topic, you should be able to:

- describe the fluid mosaic model of cell membranes;
- state that regions of hydrophobic R groups allow strong hydrophobic interactions that hold integral membrane proteins within the phospholipid bilayer;
- state that some integral membrane proteins are transmembrane proteins;
- state that peripheral membrane proteins have hydrophilic R groups on their surface and are bound to the surface of membranes, mainly by ionic and hydrogen bond interactions;
- state that many peripheral membrane proteins interact with the surfaces of integral membrane proteins;
- describe the phospholipid bilayer as a barrier to ions and most uncharged polar molecules;
- give examples of molecules which pass through the bilayer by simple diffusion;

Learning objective continued

- describe the process of facilitated diffusion;
- state that to perform specialised functions, different cell types have different channel and transporter proteins;
- state that channels are multi-subunit proteins with the subunits arranged to form water-filled pores that extend across the membrane and most channel proteins in animal and plant cells are highly selective;
- state that some channel proteins are gated and change conformation to allow or prevent diffusion;
- describe how ligand-gated channels function;
- describe how voltage-gated channels function;
- state that transporter proteins bind to the specific substance to be transported and undergo a conformational change to transfer the solute across the membrane;
- describe the process of active transport;
- state that a source of metabolic energy is required for active transport;
- state that some active transport proteins hydrolyse ATP to provide the energy fo the conformational change required to move substances across the membrane;
- describe the role of ATPases;
- define the term electrochemical gradient;
- describe the role of ion pumps;
- describe the role and function of the sodium-potassium pump;
- state that the sodium-potassium pump is found in most animal cells, accounting for a high proportion of the basal metabolic rate in many organisms;
- describe the role of the sodium potassium pump in driving the active transport of glucose in the small intestine;
- describe the role of the glucose transporter responsible for glucose symport.

TOPIC 3. MEMBRANE PROTEINS

3.1 Movement of molecules across membranes

The fluid mosaic model

The fluid mosaic model describes the structure of the plasma membrane.

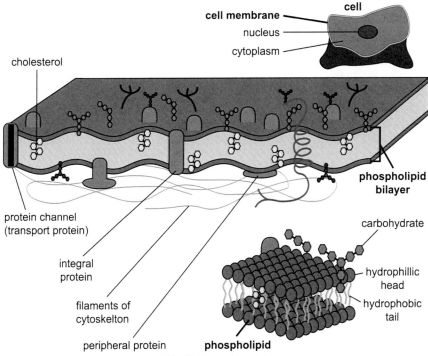

The fluid mosaic model

Membranes are comprised of a bilayer of phospholipid molecules and a patchwork of protein molecules. The head region of a phospholipid molecule is charged and, therefore, hydrophilic (attracted to water). The tail region is uncharged and non-polar, and, therefore, hydrophobic (repelled by water). The phospholipids are constantly changing position; this gives the membrane its fluid quality.

The cell membrane is embedded with proteins which form a patchy mosaic. Proteins found within the membrane can have many different functions. Membrane proteins can be classed as integral or peripheral.

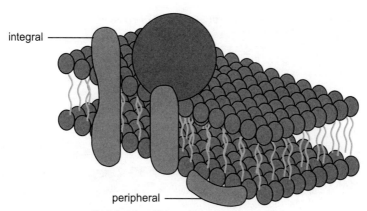

Peripheral and integral membrane proteins

Integral membrane proteins are held firmly in place within the membrane. They are held in place by strong hydrophobic interactions with the phospholipid tails. Regions of hydrophobic R groups allow strong hydrophobic interactions that hold integral membrane proteins within the phospholipid bilayer. Some integral membrane proteins are transmembrane, this means that they span the entire width of the membrane. Examples of transmembrane proteins include channels, transporters and many receptors.

Peripheral membrane proteins are only loosely associated with the plasma membrane. Peripheral membrane proteins have hydrophilic R groups on their surface and are bound to the surface of membranes, mainly by ionic and hydrogen bond interactions. Many peripheral membrane proteins interact with the surfaces of integral membrane proteins.

Facilitated diffusion

The phospholipid bilayer is a barrier to ions and most uncharged polar molecules. Some small molecules, such as oxygen and carbon dioxide, pass through the bilayer by simple diffusion, however, most molecules require membrane proteins to enter or exit the cell. Facilitated diffusion is the **passive** transport of substances across the membrane through specific transmembrane proteins. For example water can pass across the plasma membrane by diffusing through the phospholipid bilayer or through water channels which are called aquaporins. Water diffuses very slowly across the plasma membrane; most water passes across the membrane via aquaporin which can allow up to 3 billion water molecules to move across the membrane per second. The direction of water movement is dependent upon the osmotic gradient.

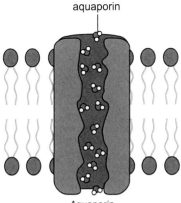

Aquaporin

To perform specialised functions, different cell types have different channel and transporter proteins. Channels are multi-subunit proteins with the subunits arranged to form water-filled pores that extend across the membrane as shown in the following diagram.

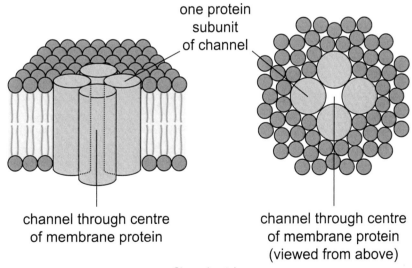

Channel proteins

Some channel proteins are gated and change **conformation** to allow or prevent diffusion. Gated channels respond to a stimulus which causes them to open or close. The stimulus may be chemical (ligand-gated) or electrical (voltage-gated).

Ligand-gated channels are controlled by the binding of signal molecules.

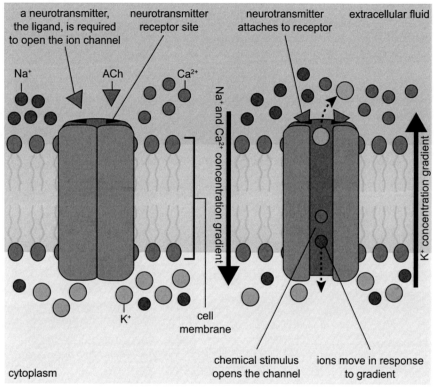

Ligand-gated channel

Voltage-gated channels are controlled by changes in ion concentrations.

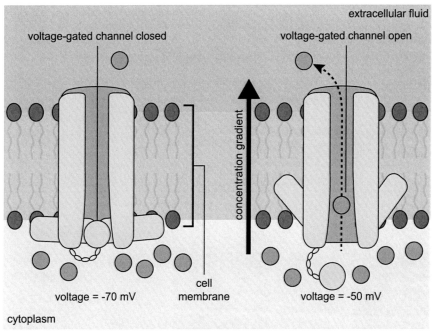

Voltage-gated channel

Transporter proteins bind to the specific substance to be transported and undergo a conformational change to transfer the solute across the membrane. The transporters actually alternate between two conformations so that the binding site for a solute is sequentially exposed on one side of the bilayer, then the other, therefore allowing the solute to move into or out of the cell. This process is illustrated in the following diagram.

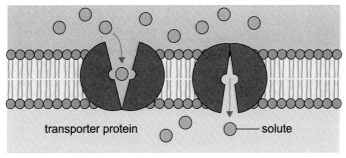

Conformational change in transporter protein

Active transport

Some transporter proteins require energy to bring about the necessary conformational change to transport a solute across the membrane. In this case, the transport is **active** (requires a source of metabolic energy). Active transport uses pump proteins that transfer substances across the membrane against their concentration gradient. The conformational change in active transport requires energy from hydrolysis of ATP.

Pumps that mediate active transport are transporter proteins coupled to an energy source. Some active transport proteins hydrolyse ATP directly to provide the energy for the conformational change required to move substances across the membrane. This is shown in the diagram below which illustrates the sodium potassium pump, also known as the NA/K ATPase (ATPases hydrolyse ATP).

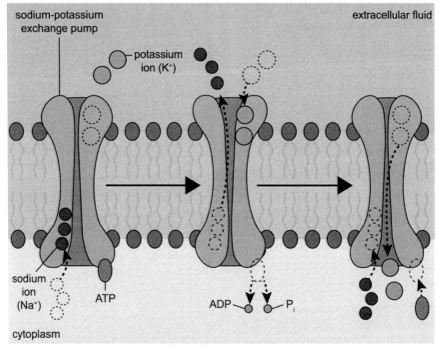

Sodium-potassium pump (Na/K-ATPase)

3.2 Ion transport pumps and generation of ion gradients

All cells have an electrical potential difference (voltage) across their plasma membrane. This voltage is called the membrane potential. In neurons, the membrane potential is typically between -60 and -80 mV (millivolts) when the cell is not transmitting signals. The minus sign means that the inside of the cell is negative relative to the outside. For a solute carrying a net charge (such as a sodium ion which carries one positive charge), the concentration gradient and the electrical potential difference combine to form the electrochemical gradient that determines the transport of the solute.

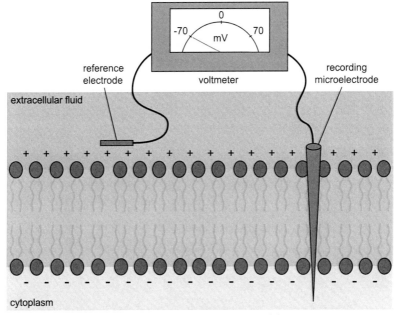

Measurement of membrane potential

Ion pumps, such as the sodium-potassium pump, use energy from the hydrolysis of ATP to establish and maintain ion gradients. The sodium-potassium pump transports ions against a steep concentration gradient using energy directly from ATP hydrolysis. It actively transports sodium ions out of the cell and potassium ions into the cell.

a) The transporter protein has high affinity for sodium ions inside the cell therefore binding occurs.

b) Phosphorylation by ATP causes the **conformation** of the protein to change.

c) The affinity for ions changes resulting in sodium being released outside of the cell.

d) Potassium ions from outside the cell bind to the sodium-potassium pump.

e) Dephosphorylation occurs which causes the conformation of the protein to change.

f) Potassium ions are taken into the cell and the affinity returns to the start.

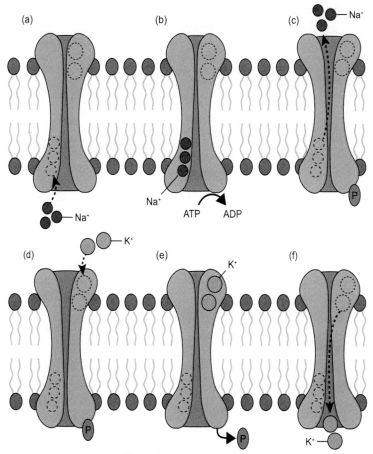

The sodium-potassium pump

For each ATP hydrolysed, three sodium ions are transported out of the cell and two potassium ions are transported into the cell. This establishes both concentration gradients and an electrical gradient. The sodium-potassium pump is found in most animal cells, accounting for a high proportion of the basal metabolic rate in many organisms.

In intestinal epithelial cells the sodium-potassium pump generates a sodium ion gradient across the plasma membrane. The sodium gradient created by the sodium-potassium pump drives the active transport of glucose. This allows glucose to be absorbed from the small intestine into the bloodstream. The glucose transporter responsible for this glucose **symport** transports sodium ions and glucose at the same time and in the same direction. Sodium ions enter the cell down their concentration gradient; the simultaneous transport of glucose pumps glucose into the cell against its concentration gradient.

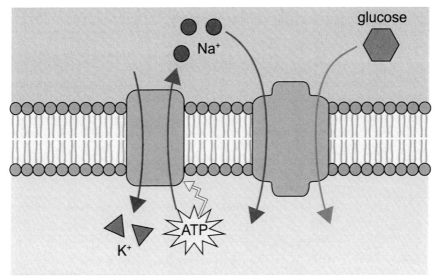

Sodium potassium pump and glucose symport

3.3 Learning points

Summary

- The fluid mosaic model describes the structure of the plasma membrane; membranes are comprised of a bilayer of phospholipid molecules and a patchwork of protein molecules.

- The head region of a phospholipid molecule is charged and, therefore, hydrophilic (attracted to water) and the tail region is uncharged and non-polar, and, therefore, hydrophobic (repelled by water).

- Membrane proteins may be integral or peripheral.

- Regions of hydrophobic R groups allow strong hydrophobic interactions that hold integral membrane proteins within the phospholipid bilayer.

- Some integral membrane proteins are transmembrane proteins meaning they span the width of the membrane.

- Integral membrane proteins interact extensively with the hydrophobic region of membrane phospholipids.

- Peripheral membrane proteins have hydrophilic R groups on their surface and are bound to the surface of membranes, mainly by ionic and hydrogen bond interactions.

> **Summary continued**

- Many peripheral membrane proteins interact with the surfaces of integral membrane proteins.
- The phospholipid bilayer is a barrier to ions and most uncharged polar molecules.
- Some small molecules, such as oxygen and carbon dioxide, pass through the bilayer by simple diffusion.
- Facilitated diffusion is the passive transport of substances across the membrane through specific transmembrane proteins.
- To perform specialised functions, different cell types have different channel and transporter proteins.
- Most channel proteins in animal and plant cells are highly selective.
- Channels are multi-subunit proteins with the subunits arranged to form water-filled pores that extend across the membrane.
- Some channel proteins are gated and change conformation to allow or prevent diffusion.
- Ligand-gated channels are controlled by the binding of signal molecules, and voltage-gated channels are controlled by changes in ion concentration.
- Transporter proteins bind to the specific substance to be transported and undergo a conformational change to transfer the solute across the membrane.
- Transporters alternate between two conformations so that the binding site for a solute is sequentially exposed on one side of the bilayer, then the other.
- Active transport uses pump proteins that transfer substances across the membrane against their concentration gradient.
- Pumps that mediate active transport are transporter proteins coupled to an energy source.
- A source of metabolic energy is required for active transport.
- Some active transport proteins hydrolyse ATP directly to provide the energy for the conformational change required to move substances across the membrane.
- ATPases hydrolyse ATP.
- For a solute carrying a net charge, the concentration gradient and the electrical potential difference combine to form the electrochemical gradient that determines the transport of the solute.
- A membrane potential (an electrical potential difference) is created when there is a difference in electrical charge on the two sides of the membrane.
- Ion pumps, such as the sodium-potassium pump, use energy from the hydrolysis of ATP to establish and maintain ion gradients.
- The sodium-potassium pump transports ions against a steep concentration gradient using energy directly from ATP hydrolysis.

Summary continued

- The pump has high affinity for sodium ions inside the cell; binding occurs; phosphorylation by ATP; conformation changes; affinity for sodium ions decreases; sodium ions released outside of the cell; potassium ions bind outside the cell; dephosphorylation; conformation changes; potassium ions taken into cell; affinity returns to start.

- For each ATP hydrolysed, three sodium ions are transported out of the cell and two potassium ions are transported into the cell; this establishes both concentration gradients and an electrical gradient.

- The sodium-potassium pump is found in most animal cells, accounting for a high proportion of the basal metabolic rate in many organisms.

- In the small intestine, the sodium gradient created by the sodium-potassium pump drives the active transport of glucose.

- In intestinal epithelial cells the sodium-potassium pump generates a sodium ion gradient across the plasma membrane.

- The glucose transporter responsible for this glucose symport transports sodium ions and glucose at the same time and in the same direction.

- Sodium ions enter the cell down their concentration gradient; the simultaneous transport of glucose pumps glucose into the cell against its concentration gradient.

3.4 Extended response question

The activity which follows presents an extended response question similar to the style that you will encounter in the examination.

You should have a good understanding of the sodium potassium pump before attempting the question.

You should give your completed answer to your teacher or tutor for marking, or try to mark it yourself using the suggested marking scheme.

Extended response question: The sodium potassium pump

Describe the action of the sodium potassium pump. *(10 marks)*

3.5 End of topic test

End of Topic 3 test — Go online

The following diagram shows a series of proteins in a section of plasma membrane.

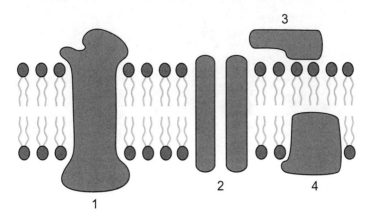

Q1: Which number(s) in the diagram represent integral protein? *(1 mark)*

....

Q2: Which number(s) in the diagram represent peripheral protein? *(1 mark)*

....

Q3: How are integral proteins held within the membrane? *(1 mark)*

....

Q4: Facilitated transport through transporter proteins is a _____ process, meaning that it does not require energy. *(1 mark)*

....

Q5: Some transporter proteins require a source of metabolic energy to bring about the necessary conformational change. In this case the transport is _____. *(1 mark)*

....

TOPIC 3. MEMBRANE PROTEINS

Q6: What type of channel is shown in the following diagram? *(1 mark)*

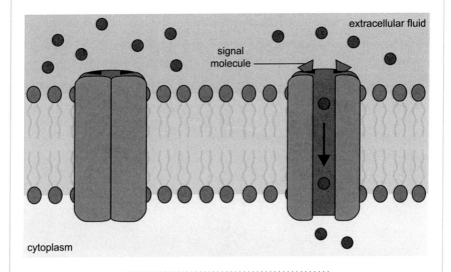

....

Q7: What name is given to the electrical potential difference (voltage) across the plasma membrane of a cell? *(1 mark)*

....

Q8: Ligand-gated ion channels allow ions to enter the cells and change the electrical potential difference across the membrane.

What type of membrane channel will open in response to a change in ion concentrations? *(1 mark)*

The following diagram shows the sodium-potassium pump.

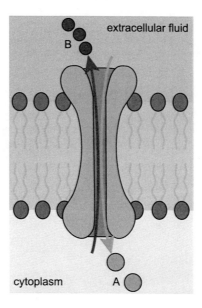

Q9: Which ions are represented by the letter A in the diagram? *(1 mark)*

a) Potassium
b) Sodium

....

Q10: Which ions are represented by the letter B in the diagram? *(1 mark)*

a) Potassium
b) Sodium

....

Q11: The sodium-potassium pump requires energy supplied by _____. *(1 mark)*

....

Q12: Explain what happens when the sodium-potassium pump becomes phosphorylated. *(2 marks)*

....

Q13: The sodium-potassium pump moves ions on the ratio 3 sodium : 2 potassium.

If 10,000 of these ions are pumped across the membrane every 10 seconds, how many sodium ions are moved across in one minute? *(1 mark)*

Unit 1 Topic 4

Communication within multicellular organisms

Contents

4.1 Coordination	86
4.2 Hydrophobic signals and control of transcription	87
4.3 Hydrophilic signals and transduction	91
4.4 Generation of a nerve impulse	96
4.5 Initiation of a nerve impulse in response to an environmental stimulus: the vertebrate eye . .	103
4.6 Learning points	106
4.7 Extended response question	109
4.8 End of topic test	110

Prerequisites

You should already know that:

- hormones are chemical messengers;

- hormones are produced by endocrine glands and travel in the bloodstream to target tissues;

- target tissues have receptors which are complementary to a specific hormone;

- pancreatic receptors respond to high blood glucose levels by causing secretion of insulin;

- insulin brings about the conversion of glucose to glycogen in the liver, decreasing blood glucose concentration.

Learning objective

By the end of this topic, you should be able to:

- state that multicellular organisms signal between cells using extracellular signalling molecules for example using steroid hormones, peptide hormones, and neurotransmitters;
- describe the role of receptor molecules;
- state that different cell types produce specific signals that can only be detected and responded to by cells with the specific receptor;
- state that signalling molecules may have different effects on different target cell types due to differences in the intracellular signalling molecules and pathways that are involved;
- state that in a multicellular organism, different cell types may show a tissue-specific response to the same signal;
- state that hydrophobic signalling molecules can diffuse directly through the phospholipid bilayers of membranes, and so bind to intracellular receptors;
- state that the receptors for hydrophobic signalling molecules are transcription factors;
- describe the role of transcription factors;
- give examples of steroid hormones which are hydrophobic signalling molecules;
- state that steroid hormones bind to specific receptors in the cytosol or the nucleus;
- state that the hormone-receptor complex moves to the nucleus where it binds to specific sites on DNA and affects gene expression;
- state that the hormone-receptor complex binds to specific DNA sequences called hormone response elements (HREs) - binding at these sites influences the rate of transcription, with each steroid hormone affecting the gene expression of many different genes;
- state that hydrophilic signalling molecules bind to transmembrane receptors and do not enter the cytosol;
- give examples of hydrophilic extracellular signalling molecules;
- describe the role of transmembrane receptors in signal transduction;
- state that transmembrane receptors act as signal transducers by converting the extracellular ligand-binding event into intracellular signals, which alters the behaviour of the cell;
- describe the roles of G-proteins and cascades of phosphorylation by kinase enzymes in transducing hydrophilic signals;
- describe the effects of the binding of the peptide hormone insulin to its receptor;
- describe the causes of type 1 and type 2 diabetes;

TOPIC 4. COMMUNICATION WITHIN MULTICELLULAR ORGANISMS 85

Learning objective continued

- state that type 2 is generally associated with obesity;
- describe the role of exercise in the treatment of type 2 diabetes;
- define the term resting membrane potential;
- state that the transmission of a nerve impulse requires changes in the membrane potential of the neuron's plasma membrane;
- define the term action potential;
- state that neurotransmitters initiate a response by binding to their receptors (ligand-gated ion channels) at a synapse;
- state that depolarisation of the plasma membrane as a result of the entry of positive ions triggers the opening of voltage-gated sodium channels, and further depolarisation occurs;
- state that inactivation of the sodium channels and the opening of potassium channels restores the resting membrane potential;
- describe how a wave of depolarisation passes along the length of a neuron;
- describe what happens when an action potential reaches the end of a neuron;
- state that restoration of the resting membrane potential allows the inactive voltage-gated sodium channels to return to a conformation that allows them to open again in response to depolarisation of the membrane;
- describe how ion concentration gradients are re-established;
- describe the role of the retina;
- state that in animals the light-sensitive molecule retinal is combined with a membrane protein, opsin, to form the photoreceptors of the eye;
- name the retinal-opsin complex found in rod cells;
- describe the process which triggers nerve impulses in neurons in the retina beginning when retinal absorbs a photon of light;
- state that a very high degree of amplification results in rod cells being able to respond to low intensities of light;
- describe the importance of different forms of opsin in cone cells.

© HERIOT-WATT UNIVERSITY

4.1 Coordination

Multicellular organisms show division of labour, this means that different cells carry out different functions within defined areas of the body.

The cells of the body must be able to communicate with each other. They must be able to receive information from other parts of the body and act upon it. Multicellular organisms achieve coordination of communication by means of **extracellular** signalling molecules. Steroid hormones, peptide hormones, and neurotransmitters are examples of extracellular signalling molecules. One tissue in the body will release a signalling molecule which will travel to another tissue with complementary receptors. This allows a signal to be passed from one cell to another.

Receptor molecules of target cells are proteins with a binding site for a signal molecule. Binding changes the **conformation** of the receptor, which initiates a response within the cell.

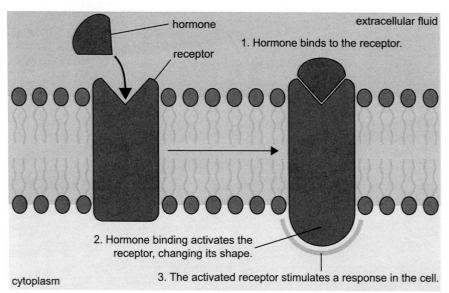

Signal reception

Different cell types produce specific signals which can only be detected and responded to by cells with the specific receptor. In a multicellular organism, different cell types may show a tissue-specific response to the same signal and signalling molecules may have different effects on different target cell types due to differences in the intracellular signalling molecules and pathways that are involved.

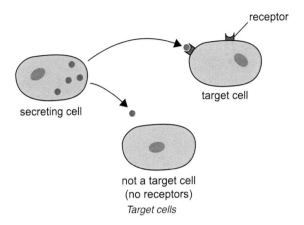

Target cells

4.2 Hydrophobic signals and control of transcription

Hydrophobic signalling molecules can diffuse directly through the phospholipid bilayers of membranes, and so bind to intracellular receptors. They are able to pass through the phospholipid bilayer because the tails of the phospholipids in the plasma membrane are also hydrophobic and allow the molecules to pass across. Hydrophobic signals can directly influence transcription of genes. The steroid hormones oestrogen and testosterone are examples of hydrophobic signalling molecules.

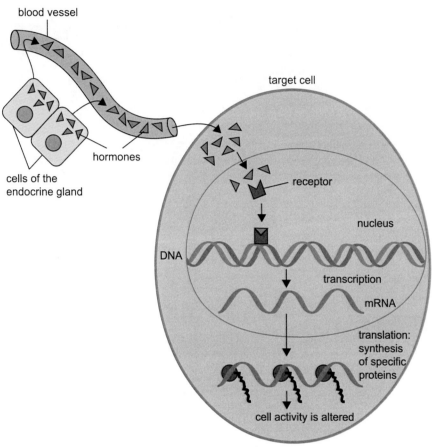

Hydrophobic signalling

The receptors for hydrophobic signalling molecules are transcription factors which can be found in the cytosol or nucleus. Transcription factors are proteins that when bound to DNA can either stimulate or inhibit initiation of transcription, they do this by enhancing or blocking the binding of RNA polymerase to specific genes (thereby controlling whether the gene is transcribed or not).

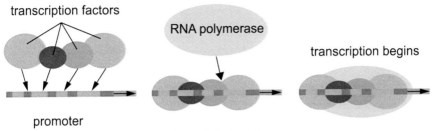

Transcription factors

When a steroid hormone enters a cell it binds to and activates a specific receptor in the cytosol (or nucleus) forming a hormone-receptor complex. The hormone-receptor complex then moves to the nucleus where it binds to specific sites on DNA called hormone response elements (HREs). Binding at these sites influences the rate of transcription, with each steroid hormone affecting the gene expression of many different genes. For example oestrogen has been found to influence the transcription of more than 80 different genes.

Steroid hormone passes across the plasma membrane

The hormone binds to the receptor protein, activating it

The hormone-receptor complex binds to specific DNA sequences called hormone response elements (HREs)

Binding at the HRE influences the rate of transcription

Steroid hormones

Cell signalling: The action of testosterone

Q1: Complete the diagram by matching the labels to the gaps.

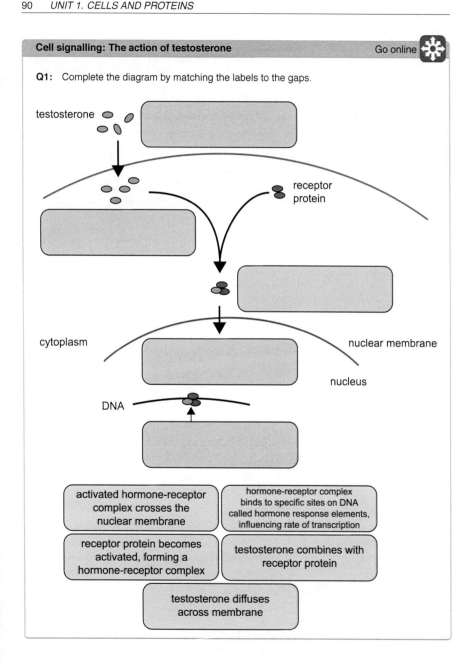

4.3 Hydrophilic signals and transduction

Hydrophilic signalling molecules

Hydrophilic signals require transmembrane receptor molecules at the surface of the cell because they are not capable of passing across the **hydrophobic** plasma membrane and entering the cytosol. Peptide hormones and neurotransmitters are examples of hydrophilic extracellular signalling molecules.

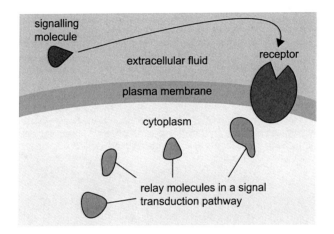

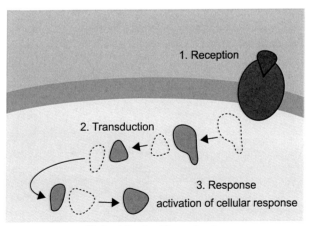

Hydrophilic signalling molecules

Stage 1 - reception

Transmembrane receptors change **conformation** (shape) when the **ligand** (signalling molecule) binds to the **extracellular** face. The signal molecule does not enter the cell, but the signal is transduced across the membrane of the cell.

Stage 2 - transduction

Transmembrane receptor proteins act as signal transducers by convert an extracellular ligand-binding event to a specific **intracellular** response through a signal transduction pathway. Transduced hydrophilic signals often involve cascades of G-proteins or **phosphorylation** by kinase enzymes.

Transduction by G-proteins

G-protein-coupled receptors (GPCRs) are linked to a G-protein. The G-protein acts as a switch that is either on or off, depending on which of the two guanine nucleotides (GDP or GTP) is attached. When a hydrophilic signalling molecule binds to the extracellular side of a GPRC, a cascade of events is initiated. Initially, when GDP is bound, the G-protein is inactive. On binding of a hydrophilic hormone to the receptor, GTP replaces GDP in the G-protein, and the G-protein becomes active. The active G-protein stimulates an enzyme, leading to a response in the cell. The response is only temporary because the G-protein also acts as a GTPase and soon hydrolyses the bound GTP into GDP, making the G-protein inactive again.

G-protein linked receptors

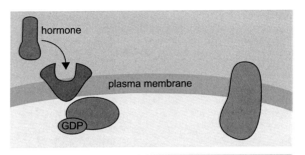

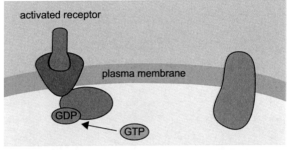

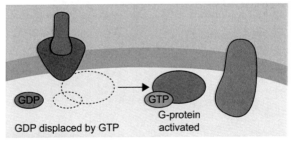

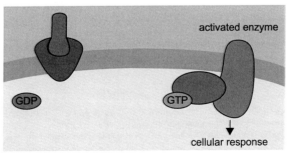

Transduction by phosphorylation

Receptor tyrosine kinases (RTKs) are transmembrane proteins. Their extracellular domain is capable of binding ligands (such as hydrophilic hormones) and their intracellular domain functions as a kinase enzyme. Kinase enzymes carry out phosphorylation reactions (addition of a phosphate group to substrates).

When a ligand binds to a RTK, tyrosine amino acids on the receptor become phosphorylated, causing a conformational change in the receptor. This results in the kinase domains in the receptors becoming activated and phosphorylating downstream cytoplasmic molecules. Phosphorylation cascades involve a series of events with one kinase activating the next in the sequence and so on. Phosphorylation cascades can result in the phosphorylation of many proteins as a result of the original signalling event. Phosphorylation cascades allow more than one intracellular signalling pathway to be activated.

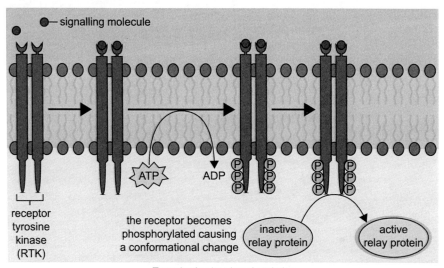

Transduction by phosphorylation

Stage 3 - response

The response of the cell will vary depending on the signal. For example, hormone binding may trigger recruitment of a channel protein to the surface of the cell. A hormone binding to a GPCR may result in the opening of an ion channel, while a hormone binding to a RTK may bring about transcription of certain genes. We are going to look at the action of two hydrophilic signalling molecules in more detail - insulin.

Insulin

The levels of glucose in the blood must be kept within strict limits. Hormones are involved in maintaining a constant blood glucose level. An increase in blood glucose concentration is detected by cells in the pancreas, which produce insulin. Insulin plays an important role in allowing fat tissue and skeletal muscles to absorb glucose from the bloodstream. Glucose passes into cells by travelling through transporter proteins in the plasma membrane (by facilitated diffusion).

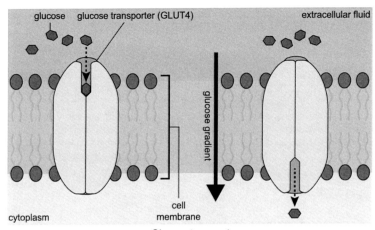

Glucose transport

Binding of the peptide hormone insulin to its receptor causes a conformational change that triggers phosphorylation of the receptor. This starts a phosphorylation cascade inside the cell, which eventually leads to GLUT4-containing vesicles being transported to the cell membrane of fat and muscle cells. GLUT4 glucose transporter proteins allow glucose to pass across the plasma membrane and enter the cell.

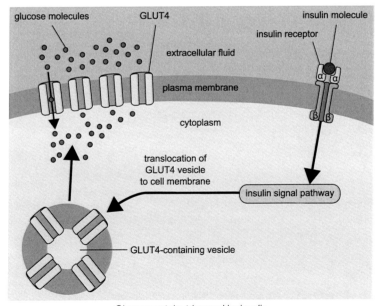

Glucose uptake triggered by insulin

In some individuals, there is a failure at some stage of the insulin signalling pathway. This results in a condition called diabetes. There are two type of diabetes.

- Type 1 - caused by a failure to produce insulin in the pancreas.
- Type 2 - caused by loss of insulin receptor function. This type of diabetes is usually associated with obesity.

Type 1 diabetes is treated with regular injections of insulin throughout the day. Type 2 diabetes may be treated with medications to lower blood glucose levels along with lifestyle changes, which may include consuming less sugar and increasing activity levels (this will aid weight loss if this is necessary). Exercise also triggers recruitment of GLUT4 so can improve uptake of glucose to fat and muscle cells in subjects with type 2 diabetes.

4.4 Generation of a nerve impulse

A membrane potential (an electrical potential difference) is created when there is a difference in electrical charge on the two sides of the membrane. Resting membrane potential is a state where there is no net flow of ions across the membrane. In neurons the resting membrane potential is typically between -60 and -80mV (millivolts).

The transmission of a nerve impulse requires changes in the membrane potential of the neuron's plasma membrane. Nerve transmission is a wave of depolarisation of the resting potential of a neuron; depolarisation is a change in the membrane potential to a less negative value inside. Depolarisation results from a wave of electrical excitation along a neuron's plasma membrane called an action potential.

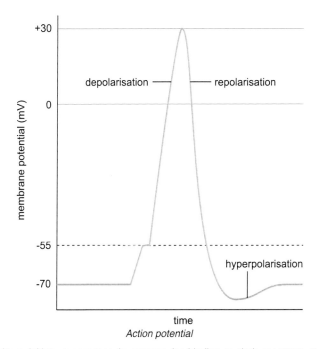
Action potential

Neurotransmitters initiate a response in neurons by binding to their receptors at a synapse. Neurotransmitter receptors are ligand-gated ion channels which open in response to binding and allow entry of positively charged ions and therefore cause depolarisation of the plasma membrane.

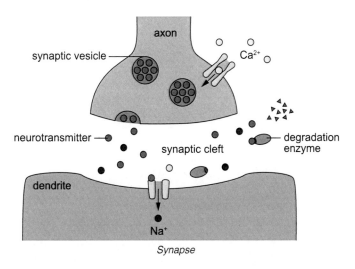
Synapse

If sufficient ion movement occurs, and the membrane is depolarised beyond a threshold value, the opening of voltage-gated sodium channels is triggered and sodium ions enter the cell down their electrochemical gradient. This leads to further depolarisation as a result of a rapid and large change in the membrane potential.

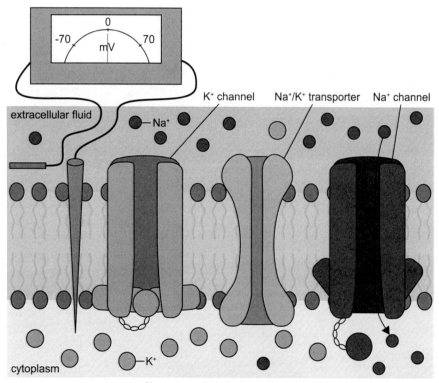

Nerve transmission - depolarisation

A short time after opening, the sodium channels become inactivated. Voltage-gated potassium channels then open to allow potassium ions to move out of the cell to restore the resting membrane potential.

TOPIC 4. COMMUNICATION WITHIN MULTICELLULAR ORGANISMS

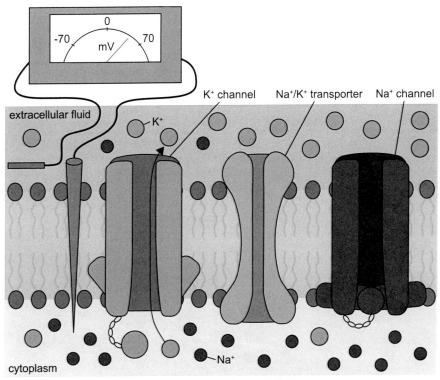

Nerve transmission - hyperpolarisation

Restoration of the resting membrane potential allows the inactive voltage-gated sodium channels to return to a conformation that allows them to open again in response to depolarisation of the membrane. Following repolarisation the sodium and potassium ion concentration gradients are reduced. The sodium-potassium pump restores the sodium and potassium ions back to resting potential levels by actively transporting excess ions in and out of the cell.

UNIT 1. CELLS AND PROTEINS

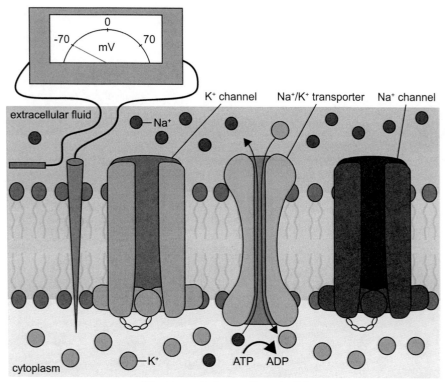

Nerve transmission - resting potential

Summary:

1. Binding of a neurotransmitter triggers the opening of ligand-gated ion channels at a synapse.

2. Ion movement occurs and there is depolarisation of the plasma membrane.

3. If sufficient ion movement occurs, and the membrane is depolarised beyond a threshold value, the opening of voltage-gated sodium channels is triggered and sodium ions enter the cell down their electrochemical gradient.

4. This leads to a rapid and large change in the membrane potential.

5. A short time after opening, the sodium channels become inactivated.

6. Voltage-gated potassium channels then open to allow potassium ions to move out of the cell to restore the resting membrane potential.

7. Ion concentration gradients are re-established by the sodium-potassium pump, which actively transports excess ions in and out of the cell.

TOPIC 4. COMMUNICATION WITHIN MULTICELLULAR ORGANISMS

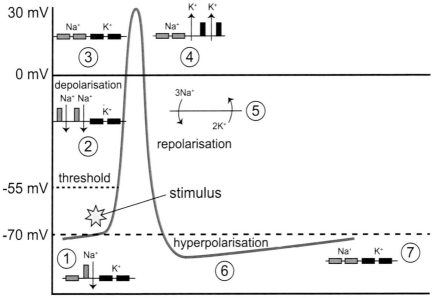

Events during an action potential

Depolarisation of a patch of membrane causes neighbouring regions of membrane to depolarise and go through the same cycle, as adjacent voltage-gated sodium channels are opened.

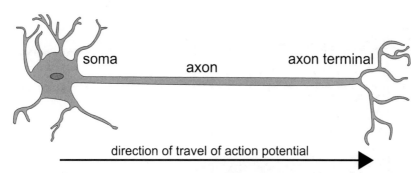

a. In response to a signal, the soma end of the axon becomes depolarised.

b. The depolarisation spreads down the axon. Meanwhile, the first part of the membrane repolarises. Because Na$^+$ channels are inactivated and additional K$^+$ channels have opened, the membrane cannot depolarise again.

c. The action potential continues to travel down the axon.

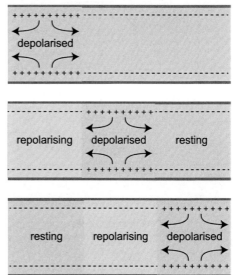

Passage of an action potential

When the action potential reaches the end of the neuron it causes vesicles containing neurotransmitter to fuse with the membrane — this releases neurotransmitter into the synapse, which stimulates a response in a connecting cell.

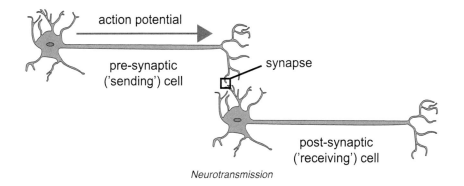

Neurotransmission

4.5 Initiation of a nerve impulse in response to an environmental stimulus: the vertebrate eye

The following diagram shows the structure of the eye. This section will focus on two types of cells found in the retina (the tissue at the back of the eye which converts light into electrical signals).

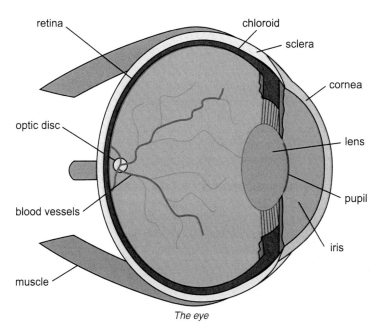

The eye

Animals have two types of photoreceptor cells within the retina of the eye; about 125 million rods and 6 million cones (named due to their shapes). These cells are shown in the following diagram.

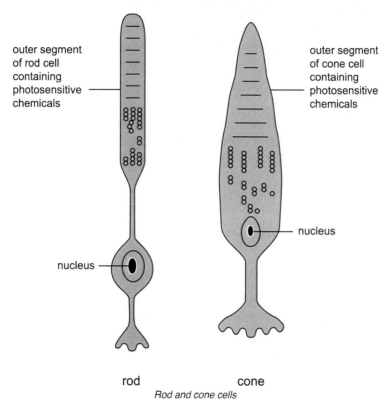

Rod and cone cells

Rod cells contain one type of light-sensitive pigment. These cells are sensitive to changes in light intensity and are particularly useful for vision in areas of low light intensity (e.g. a dim room) but do not allow colour perception. Nocturnal animals have a greater proportion of rod cells in their retina which gives them better vision at night.

Cone cells are not as sensitive to light as rod cells; however, they are particularly sensitive to specific colours (wavelengths) of light: green, red, blue and (in some animals) UV. Cone cells allow animals to have colour vision and only function in bright light. People who are colour blind lack a particular type of cone cell in their retina.

Each rod cell in the retina contains photoreceptor proteins that consist of a light-absorbing molecule, called retinal, that is bonded to a membrane protein, called **opsin**. In combination, opsin and retinal make up the visual pigment rhodopsin (as shown in the following diagram). In rod cells, the rhodopsin absorbs a wide range of wavelengths, and a very high degree of amplification from a single photon of light results in sensitivity at low light intensities. In cone cells, different forms of opsin combine with retinal to give photoreceptor proteins, each with maximal sensitivity to specific wavelengths (red, green, blue or UV).

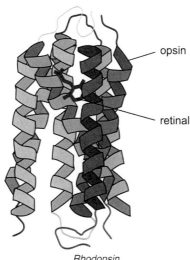

Rhodopsin

Generation of a nerve impulse is brought about when:

- **retinal** absorbs a **photon** of light and rhodopsin changes conformation to photoexcited rhodopsin;
- photoexcited rhodopsin activates a **G-protein** called transducin (a single photoexcited rhodopsin activates hundreds of molecules of transducin);
- transducin activates the enzyme phosphodiesterase (PDE) (each activated molecule of transducin activates one molecule of PDE);
- PDE catalyses the hydrolysis of a molecule called cyclic GMP (cGMP) (each active PDE molecule breaks down thousands of cGMP molecules per second);
- the reduction in cGMP concentration as a result of its hydrolysis causes the closure of ion channels in the membrane of the rod cells;
- the inward leakage of positive ions (Na^+ and Ca^+) is halted so the membrane potential increases; **hyperpolarisation** (increasing charge) triggers nerve impulses in neurons in the retina.

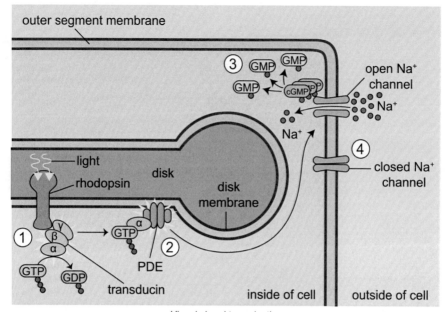

Visual signal transduction

1. Light stimulation of rhodopsin leads to activation of a G-protein, transducin.
2. Activated G-protein activates cGMP phosphodiesterase (PDE).
3. PDE hydrolyses cGMP, reducing its concentration.
4. This leads to closure of Na^+ channels.

4.6 Learning points

Summary

- Multicellular organisms signal between cells using extracellular signalling molecules.
- Steroid hormones, peptide hormones, and neurotransmitters are examples of extracellular signalling molecules.
- Receptor molecules of target cells are proteins with a binding site for a specific signal molecule.
- Binding changes the conformation of the receptor, which initiates a response within the cell.
- Different cell types produce specific signals that can only be detected and responded to

Summary continued

- by cells with the specific receptor.
- In a multicellular organism, different cell types may show a tissue-specific response to the same signal.
- Signalling molecules may have different effects on different target cell types due to differences in the intracellular signalling molecules and pathways that are involved.
- Hydrophobic signalling molecules can diffuse directly through the phospholipid bilayers of membranes, and so bind to intracellular receptors.
- The receptors for hydrophobic signalling molecules are transcription factors.
- Transcription factors are proteins that when bound to DNA can either stimulate or inhibit initiation of transcription.
- The steroid hormones oestrogen and testosterone are examples of hydrophobic signalling molecules.
- Steroid hormones bind to specific receptors in the cytosol or the nucleus; the hormone-receptor complex moves to the nucleus where it binds to specific sites on DNA and affects gene expression.
- The hormone-receptor complex binds to specific DNA sequences called hormone response elements (HREs). Binding at these sites influences the rate of transcription, with each steroid hormone affecting the gene expression of many different genes.
- Hydrophilic signalling molecules bind to transmembrane receptors and do not enter the cytosol.
- Transmembrane receptors change conformation when the ligand binds to the extracellular face; the signal molecule does not enter the cell, but the signal is transduced across the plasma membrane.
- Peptide hormones and neurotransmitters are examples of hydrophilic extracellular signalling molecules.
- Transmembrane receptors act as signal transducers by converting the extracellular ligand-binding event into intracellular signals, which alters the behaviour of the cell.
- Transduced hydrophilic signals often involve G-proteins or cascades of phosphorylation by kinase enzymes.
- G-proteins relay signals from activated receptors (receptors that have bound a signalling molecule) to target proteins such as enzymes and ion channels.
- Phosphorylation cascades allow more than one intracellular signalling pathway to be activated.
- Phosphorylation cascades involve a series of events with one kinase activating the next in the sequence and so on. Phosphorylation cascades can result in the phosphorylation of many proteins as a result of the original signalling event.

Summary continued

- Binding of the peptide hormone insulin to its receptor causes a conformational change that triggers phosphorylation of the receptor. This starts a phosphorylation cascade inside the cell, which eventually leads to GLUT4 (glucose transporter proteins) containing vesicles being transported to the cell membrane of fat and muscle cells.

- Diabetes mellitus can be caused by failure to produce insulin (type 1) or loss of receptor function (type 2).

- Type 2 is generally associated with obesity.

- Exercise also triggers recruitment of GLUT4, so can improve uptake of glucose to fat and muscle cells in subjects with type 2.

- Resting membrane potential is a state where there is no net flow of ions across the membrane.

- The transmission of a nerve impulse requires changes in the membrane potential of the neuron's plasma membrane.

- An action potential is a wave of electrical excitation along a neuron's plasma membrane.

- Binding of a neurotransmitter to its receptor triggers the opening of ligand-gated ion channels at a synapse. Positive ions enter the cell and there is depolarisation of the plasma membrane. If sufficient ion movement occurs, and the membrane is depolarised beyond a threshold value, the opening of voltage-gated sodium channels is triggered and sodium ions enter the cell down their electrochemical gradient. This leads to a rapid and large change in the membrane potential (further depolarisation). A short time after opening, the sodium channels become inactivated. Voltage-gated potassium channels then open to allow potassium ions to move out of the cell to restore the resting membrane potential.

- Depolarisation of a patch of membrane causes neighbouring regions of membrane to depolarise and go through the same cycle, as adjacent voltage-gated sodium channels are opened.

- When the action potential reaches the end of the neuron it causes vesicles containing neurotransmitter to fuse with the membrane —this releases neurotransmitter, which stimulates a response in a connecting cell.

- Restoration of the resting membrane potential allows the inactive voltage-gated sodium channels to return to a conformation that allows them to open again in response to depolarisation of the membrane.

- Following repolarisation the sodium and potassium ion concentration gradients are reduced. The sodium-potassium pump restores the sodium and potassium ions back to resting potential levels.

- The retina is the area within the eye that detects light and contains two types of photoreceptor cells: rods and cones.

- In animals the light-sensitive molecule retinal is combined with a membrane protein, opsin, to form the photoreceptors of the eye.

TOPIC 4. COMMUNICATION WITHIN MULTICELLULAR ORGANISMS

Summary continued

- Rods function in dim light but do not allow colour perception.
- Cones are responsible for colour vision and only function in bright light.
- In rod cells the retinal-opsin complex is called rhodopsin.
- Retinal absorbs a photon of light and rhodopsin changes conformation to photoexcited rhodopsin; a single photoexcited rhodopsin activates hundreds of molecules of G-protein.
- Photoexcited rhodopsin activates a G-protein, called transducin, which activates the enzyme phosphodiesterase (PDE). Each activated G-protein activates one molecule of PDE.
- PDE catalyses the hydrolysis of a molecule called cyclic GMP (cGMP). Each active PDE molecule breaks down thousands of cGMP molecules per second.
- The reduction in cGMP concentration as a result of its hydrolysis results in the closure of ion channels in the membrane of the rod cells, which triggers nerve impulses in neurons in the retina.
- A very high degree of amplification results in rod cells being able to respond to low intensities of light.
- In cone cells, different forms of opsin combine with retinal to give different photoreceptor proteins, each with a maximal sensitivity to specific wavelengths: red, green, blue or UV.

4.7 Extended response question

The activity which follows presents an extended response question similar to the style that you will encounter in the examination.

You should have a good understanding of cell signalling before attempting the question.

You should give your completed answer to your teacher or tutor for marking, or try to mark it yourself using the suggested marking scheme.

Extended response question: Cell signalling

Give an account of cell signalling. *(10 marks)*

4.8 End of topic test

End of Topic 4 test — Go online

Q2: Where are the receptors for hydrophobic signalling molecules located? *(1 mark)*

Q3: Name the specific DNA sequences to which a hormone-receptor complex binds to affect transcription of a gene. *(1 mark)*

Q4: Where are the receptors for hydrophilic signalling molecules located? *(1 mark)*

Q5: To which group of hormones does oestrogen belong? *(1 mark)*

Q6: Describe the effect of insulin binding to its receptor. *(2 marks)*

Q7: Define the term 'resting membrane potential'? *(1 mark)*

Q8: What name is given to a wave of electrical excitation along a neuron's plasma membrane? *(1 mark)*

Q9: Put the following steps in order to correctly describe the generation of a nerve impulse. *(6 marks)*

- The sodium channels become inactivated.
- Voltage-gated potassium channels open to allow potassium ions to move out of the cell.
- Neurotransmitter binds to its receptor triggering the opening of ligand-gated ion channels.
- Ion movement occurs and there is depolarisation of the plasma membrane.
- This leads to a rapid and large change in the membrane potential.
- Voltage-gated sodium channels open and sodium ions enter the cell down their electrochemical gradient.

Q10: Name the membrane protein which re-establishes ion concentration gradients following repolarisation of the neuronal membrane. *(1 mark)*

Q11: Tetrodotoxin (TTX) is a potent neurotoxin found in species such as the puffer fish. TTX binds to the voltage-gated sodium channels in the membrane of neurons and inhibits their action.

Predict the effect this will have on the neuron. *(2 marks)*

The following diagram shows the structure of rhodopsin.

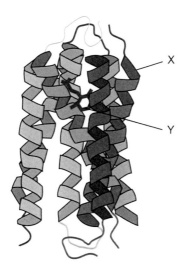

Q12: What does the letter X represent in the diagram? *(1 mark)*

....

Q13: What does the letter Y represent in the diagram? *(1 mark)*

....

Q14: What type of cell contains rhodopsin, which absorbs a wide range of wavelengths and provides sensitivity at low light intensities? *(1 mark)*

....

Q15: What type of molecule is activated by photoexcited rhodopsin? *(1 mark)*

....

Q16: Name the enzyme which catalyses the hydrolysis of cyclic GMP (cGMP). *(1 mark)*

Unit 1 Topic 5

Protein control of cell division

Contents

5.1 The cytoskeleton and cell division 115
5.2 The cell cycle 117
5.3 Control of the cell cycle 119
5.4 Control of programmed cell death (apoptosis) 122
5.5 Learning points 125
5.6 Extended response questions 127
5.7 End of topic test 128

Prerequisites

You should already know that:

- mitosis is required for growth and repair;
- the sequence of events of mitosis (including the terms chromatids, equator and spindle fibres);
- diploid cells have two matching sets of chromosomes, which are replicated during mitosis.

Learning objective

By the end of this topic, you should be able to:

- describe the role of the cytoskeleton;
- describe the structure of the cytoskeleton;
- state that microtubules control the movement of membrane-bound organelles and chromosomes;
- state that cell division requires remodelling of the cytoskeleton;
- describe the formation and breakdown of microtubules in terms of polymerisation and depolymerisation of tubulin;
- describe the role of microtubules in cell division;

Learning objective continued

- state that the cell cycle consists of interphase and mitotic (M) phase;
- name the three stages of interphase and describe the processes which take place during each stage;
- state that the mitotic phase involves mitosis and cytokinesis;
- state that mitosis consists of prophase, metaphase, anaphase and telophase and describe the events taking place in each stage;
- state that progression through the cell cycle is controlled by checkpoints;
- describe the role of checkpoints;
- describe the role of cyclin proteins in regulating the cell cycle;
- describe the role of retinoblastoma protein (Rb) at the G1 checkpoint;
- state that at the G2 checkpoint, the success of DNA replication and any damage to DNA is assessed;
- state that DNA damage triggers the activation of several proteins including p53 that can stimulate DNA repair, arrest the cell cycle or cause cell death;
- describe the role of the metaphase checkpoint;
- state that an uncontrolled reduction in the rate of the cell cycle may result in degenerative disease and an uncontrolled increase in the rate of the cell cycle may result in tumour formation;
- state that a proto-oncogene is a normal gene, usually involved in the control of cell growth or division, which can mutate to form a tumour-promoting oncogene;
- state that apoptosis is triggered by cell death signals that can be external or internal;
- give an example of an external death signal and an internal death signal;
- describe how external death signal molecules bring about apoptosis;
- describe how internal death signal molecules bring about apoptosis;
- state that both types of death signal result in the activation of caspases (types of protease enzyme) that cause the destruction of the cell;
- explain why apoptosis is essential during development of an organism;
- state that cells may initiate apoptosis in the absence of growth factors.

5.1 The cytoskeleton and cell division

The **eukaryotic** cell is a three-dimensional structure. It has a network of proteins extending throughout the cytoplasm, known as the cytoskeleton. The cytoskeleton is anchored to proteins in the plasma membrane and is dynamic in nature, constantly breaking down and re-forming. The function of the cytoskeleton is to provide mechanical support so that the cell maintains its shape.

The following image shows two proteins of the cytoskeleton. Microtubules immediately surround the small dark oval nuclei while actin filaments are found towards the edges.

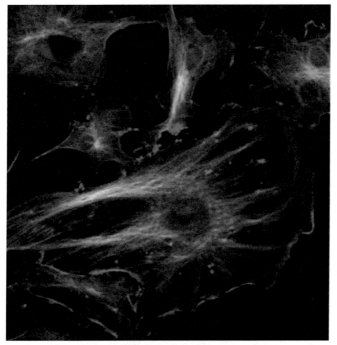

The cytoskeleton

The cytoskeleton is made up of different types of protein including microtubules. Microtubules are hollow, straight cylinders composed of proteins called tubulins. The microtubule itself is made up of alternating dimers of α and β tubulin. The formation and breakdown of microtubules involves polymerisation (growth of the microtubule) and depolymerisation (shrinkage of the microtubule) of tubulin. Microtubules control the movement of membrane-bound organelles and chromosomes.

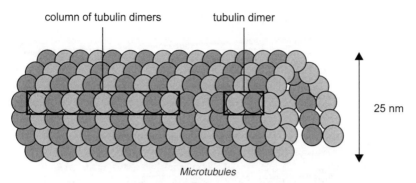

Microtubules

Microtubules are found in all eukaryotic cells and originate from the microtubule organising centre (MTOC) or centrosome. The centrosome/MTOC is found near the nucleus and contains centrioles, which are the site of microtubule synthesis within the centrosome. Microtubules play an important role in cell division as this process requires remodelling of the cell's cytoskeleton. Microtubules also form the spindle fibres, which are active during cell division. The following image shows a cell which is dividing; the microtubules of the spindle fibres are marked green and the chromosomes are marked blue.

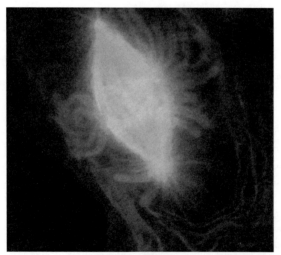

A cell undergoing mitosis

5.2 The cell cycle

Cell division allows organisms to grow and develop, to replace dead cells and repair tissue. As the term 'cell cycle' suggests, this is a continual process, but it can be divided up into distinct stages. The whole cycle can be divided up into two parts.

1. Interphase.
2. The mitotic phase.

Interphase lasts much longer than the mitotic phase. When we look at a group of cells by light microscopy, only a small proportion of them are in the mitotic phase; most of them appear to be doing nothing. Although we cannot see anything taking place, describing interphase as the 'resting phase' is in fact far from reality. Interphase is an active period of growth. During interphase, protein synthesis takes place, cytoplasmic organelles are synthesised, the cell grows and replicates its chromosomes.

Interphase is divided into three sub-phases.

1. G1 is the first 'growth' phase; it is a growth period where proteins and organelles are synthesised.
2. During S phase the DNA is replicated in preparation for mitosis.
3. The final phase is G2 (the second 'growth' phase); this is another growth period during which proteins and organelles are synthesised.

At the end of G2, cells enter the mitotic phase (M), which is divided up into two stages.

1. Mitosis - when the nucleus and its contents divide.
2. Cytokinesis - the separation of the cytoplasm into daughter cells.

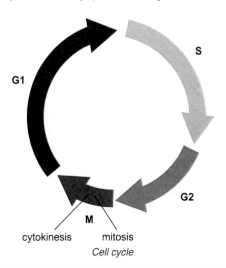

Cell cycle

Mitosis

In mitosis, the chromosomal material is separated by the spindle microtubules. This is followed by cytokinesis, in which the cytoplasm is separated into two daughter cells. Mitosis consists of prophase, metaphase, anaphase and telophase.

1. Prophase
    - DNA condenses into chromosomes each consisting of two sister chromatids.
    - Nuclear membrane breaks down; spindle microtubules extend from the MTOC by polymerisation and attach to chromosomes via their **kinetochores** in the centromere region.
    - At the end of prophase, cells enter metaphase.
2. Metaphase
    - Chromosomes are aligned at the metaphase plate (the term metaphose plate is used to describe the arrangement of the chromosomes at the equator of the cell).
    - At the end of metaphase, cells enter anaphase.
3. Anaphase
    - As spindle microtubules shorten by depolymerisation, sister chromatids are separated, and the chromosomes are pulled to opposite poles.
    - At the end of anaphase, cells enter telophase.
4. Telophase
    - The chromosomes decondense and nuclear membranes are formed around them.
    - Cytokinesis also takes place during this period, which involves the separation of the cytoplasm into two daughter cells.

Mitosis   Go online

Q1: Complete the diagram using the labels provided.

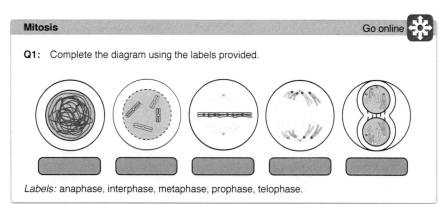

Labels: anaphase, interphase, metaphase, prophase, telophase.

Microtubules play an important role throughout this process. The spindle begins to form at prophase and is organised by the centrosomes at the two poles of the cell. The spindle fibres allow both the alignment of the chromosomes at the metaphase plate and the separation of chromatids to opposite poles. The spindle fibres also play an important role in the formation of daughter nuclei, allowing the separated chromatids to group at the opposite poles of the cell.

TOPIC 5. PROTEIN CONTROL OF CELL DIVISION

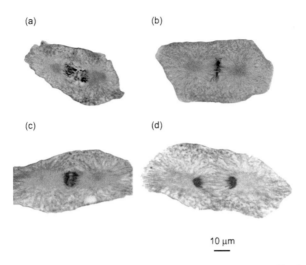

Phases of mitosis; (a) prophase; (b) metaphase; (c) early anaphase; (d) telophase

5.3 Control of the cell cycle

Cell cycle checkpoints

The timing and rate of cell division is crucial to normal growth, development and maintenance of tissues. Frequency of division will vary with cell type. For example, human skin cells divide throughout life, whereas liver cells will only divide if the need arises (for example, as a result of injury). In a mature adult, nerve cells do not divide at all.

The cell cycle must be controlled to ensure that events in the cell cycle proceed in the correct order and that each event is completed before the next starts. For example, it would be catastrophic for the cell if it entered M phase before the DNA had replicated in S phase. Sufficient nutrients and growth factors must also be present before the cell commits itself to the next stage in the cell cycle.

There are checkpoints at various stages within the cell cycle. This is where stop and start signals regulate the cycle. Checkpoints are mechanisms within the cell that assess the condition of the cell during the cell cycle and halt progression to the next phase until certain requirements are met. There are three major checkpoints in the cell cycle.

G1 checkpoint

The first checkpoint (G1 checkpoint) occurs towards the end of G1. Sufficient cell growth must have occurred and other conditions must be satisfied before the cell can enter S phase. Cyclin proteins that accumulate during cell growth are involved in regulating the cell cycle. Cyclins combine with and activate cyclin-dependent kinases (CDKs). Active cyclin-CDK complexes phosphorylate proteins that regulate progression through the cycle. If a sufficient threshold of phosphorylation is reached, the cell cycle moves on to the next stage. If an insufficient threshold is reached, the cell is held at a checkpoint.

At the G1 checkpoint, retinoblastoma protein (Rb) acts as a tumour suppressor by inhibiting the transcription of genes that code for proteins needed for DNA replication. The non-phosphorylated form of Rb restricts progression from G1 phase into S phase by binding to a transcription factor, therefore preventing transcription of certain genes required for S phase to begin; thus the cell remains in G1. Phosphorylation by G1 cyclin-CDK inhibits the activity of Rb, meaning that it can no longer bind the transcription factor. This allows transcription of the genes that code for proteins needed for DNA replication and allows cells to progress from G1 to S phase.

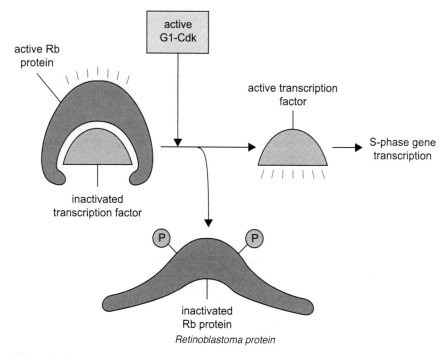

Retinoblastoma protein

G2 checkpoint

At the G2 checkpoint, the success of DNA replication and any damage to DNA is assessed. DNA damage triggers the activation of several proteins that can stimulate DNA repair, arrest the cell cycle or cause cell death. One of the proteins activated as a result of DNA damage is p53. The p53 protein has been described as 'the guardian of the genome' because of its role in maintaining a functional genome. The fact that the majority of human cancers can be linked to mutations in the p53 gene demonstrates the importance of this protein in regulating cell cycle.

Upon recognising damage to the DNA of the cell, p53 can bring about several cellular responses:

- it can activate DNA repair proteins to repair the DNA damage;
- it can arrest the cell cycle at the G1 checkpoint, which means that the cell cycle halts at this point - this can allow DNA repair proteins time to recognise and fix the DNA damage so the cell can restart the cell cycle;

- if the DNA damage is too severe, it can initiate apoptosis (programmed cell death).

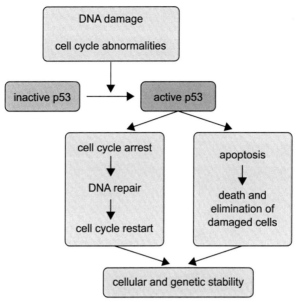

The action of p53

Metaphase checkpoint

A metaphase checkpoint controls progression from metaphase to anaphase. At the metaphase checkpoint, progression is halted until the chromosomes are aligned correctly on the metaphase plate and attached to the spindle microtubules. This checkpoint ensures that each daughter cell receives the correct number of chromosomes.

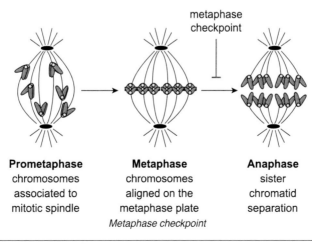

Prometaphase
chromosomes associated to mitotic spindle

Metaphase
chromosomes aligned on the metaphase plate

Metaphase checkpoint

Anaphase
sister chromatid separation

The importance of controlling cell cycle

It is important that the rate of cell cycle is tightly controlled. An uncontrolled reduction in the rate of the cell cycle may result in degenerative disease. For example, alteration in the normal control of cell cycle is thought to lead to expression of certain proteins associated with Alzheimer's disease, eventually resulting in neuronal cell death. An uncontrolled increase in the rate of the cell cycle may result in tumour formation.

A proto-oncogene is a normal gene, usually involved in the control of cell growth or division, which can mutate to form a tumour-promoting oncogene. These oncogenes cause the cell to divide in an uncontrolled and unregulated manner, for example a cell may continue to grow and divide despite the absence of growth signals. Mutation of just one proto-oncogene to an oncogene within a cell is enough to cause unregulated growth.

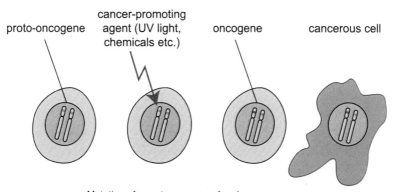

Mutation of a proto-oncogene forming an oncogene

5.4 Control of programmed cell death (apoptosis)

The destruction of cells must be carefully controlled in a multicellular organism. The destruction of cells is brought about by a process known as programmed cell death or **apoptosis**.

TOPIC 5. PROTEIN CONTROL OF CELL DIVISION

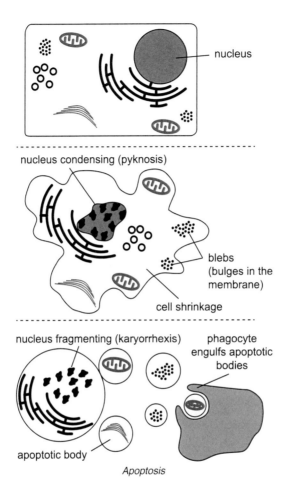

Apoptosis

Apoptosis is essential during development of an organism to remove cells no longer required as development progresses or during metamorphosis. For example, during human embryonic development, programmed cell death of the cells between the fingers and toes allows individual digits to form (rather than webbed digits). Apoptosis is also important in those organisms which undergo metamorphosis such as the development of an adult frog from a tadpole.

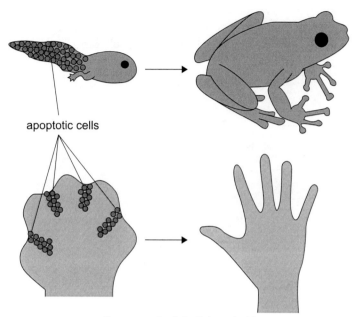

Programmed cell death (apoptosis)

Apoptosis is triggered by cell death signals that can be external or internal.

The production of death signal molecules from **lymphocytes** is an example of an external death signal. Cytotoxic T-lymphocytes express a death activator signal molecule on their surface called Fas. When a cell death signal molecule such as Fas binds to its surface receptor protein on a target cell, it activates a protein cascade that produces active **caspases** (types of protease enzyme). The activated caspases then cause the destruction of the cell.

DNA damage is an example of an internal death signal. An internal death signal resulting from DNA damage causes activation of the p53 tumour-suppressor protein. This results in the activation of caspases that cause the destruction of the cell. Cells may also initiate apoptosis in the absence of growth factors.

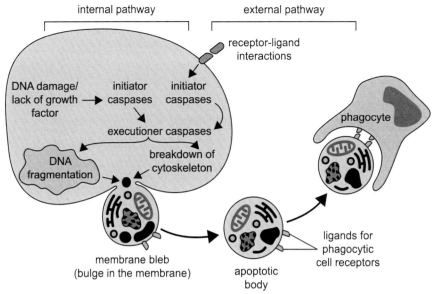
Internal and external apoptosis pathways

5.5 Learning points

Summary

- The cytoskeleton gives mechanical support and shape to cells.
- The cytoskeleton consists of different protein structures including microtubules, which are found in all eukaryotic cells.
- Microtubules are hollow cylinders composed of the protein tubulin; they radiate from the microtubule organising centre (MTOC) or centrosome.
- Microtubules control the movement of membrane-bound organelles and chromosomes.
- Cell division requires remodelling of the cytoskeleton.
- Formation and breakdown of microtubules involves polymerisation and depolymerisation of tubulin.
- Microtubules form the spindle fibres that are active during cell division.
- The cell cycle consists of interphase and mitotic (M) phase.
- Interphase involves growth and DNA synthesis including G1, a growth phase; S phase, during which the DNA is replicated; and G2, a further growth phase.

> **Summary continued**
>
> - Mitotic phase involves mitosis and cytokinesis.
> - In mitosis the chromosomal material is separated by the spindle microtubules; this is followed by cytokinesis, in which the cytoplasm is separated into two daughter cells.
> - Mitosis consists of prophase, metaphase, anaphase and telophase.
> - Prophase - DNA condenses into chromosomes each consisting of two sister chromatids. Nuclear membrane breaks down; spindle microtubules extend from the MTOC by polymerisation and attach to chromosomes via their kinetochores in the centromere region.
> - Metaphase - chromosomes are aligned at the metaphase plate (equator of the spindle).
> - Anaphase - as spindle microtubules shorten by depolymerisation, sister chromatids are separated, and the chromosomes are pulled to opposite poles.
> - Telophase - the chromosomes decondense and nuclear membranes are formed around them.
> - Progression through the cell cycle is controlled by checkpoints.
> - Checkpoints are mechanisms within the cell that assess the condition of the cell during the cell cycle and halt progression to the next phase until certain requirements are met.
> - Cyclin proteins that accumulate during cell growth are involved in regulating the cell cycle.
> - Cyclins combine with and activate cyclin-dependent kinases (CDKs). Active cyclin-CDK complexes phosphorylate proteins that regulate progression through the cycle. If sufficient phosphorylation is reached, progression occurs.
> - At the G1 checkpoint, retinoblastoma protein (Rb) acts as a tumour suppressor by inhibiting the transcription of genes that code for proteins needed for DNA replication.
> - Phosphorylation by G1 cyclin-CDK inhibits the retinoblastoma protein (Rb); this allows transcription of the genes that code for proteins needed for DNA replication and the cell progress from G1 to S phase.
> - At the G2 checkpoint, the success of DNA replication and any damage to DNA is assessed.
> - DNA damage triggers the activation of several proteins including p53 that can stimulate DNA repair, arrest the cell cycle or cause cell death.
> - A metaphase checkpoint controls progression from metaphase to anaphase.
> - At the metaphase checkpoint, progression is halted until the chromosomes are aligned correctly on the metaphase plate and attached to the spindle microtubules.
> - An uncontrolled reduction in the rate of the cell cycle may result in degenerative disease.
> - An uncontrolled increase in the rate of the cell cycle may result in tumour formation.

TOPIC 5. PROTEIN CONTROL OF CELL DIVISION

> **Summary continued**
>
> - A proto-oncogene is a normal gene, usually involved in the control of cell growth or division, which can mutate to form a tumour-promoting oncogene.
> - Apoptosis is triggered by cell death signals that can be external or internal.
> - The production of death signal molecules from lymphocytes is an example of an external death signal.
> - External death signal molecules bind to a surface receptor protein and trigger a protein cascade within the cytoplasm.
> - DNA damage is an example of an internal death signal.
> - An internal death signal resulting from DNA damage causes activation of p53 tumour-suppressor protein.
> - Both types of death signal result in the activation of caspases (types of protease enzyme) that cause the destruction of the cell.
> - Cells may initiate apoptosis in the absence of growth factors.
> - Apoptosis is essential during development of an organism to remove cells no longer required as development progresses or during metamorphosis.

5.6 Extended response questions

The activity which follows presents an extended response question similar to the style that you will encounter in the examination.

You should have a good understanding of the cell cycle and apoptosis before attempting these questions.

You should give your completed answer to your teacher or tutor for marking, or try to mark it yourself using the suggested marking scheme.

Extended response question: The cell cycle

Describe the cell cycle. *(10 marks)*

Extended response question: Apoptosis

Describe the control of apoptosis. *(5 marks)*

5.7 End of topic test

End of Topic 5 test

Q2: The cytoskeleton is composed of: *(1 mark)*

a) carbohydrate.
b) lipid.
c) nucleic acid.
d) protein.

....

Q3: All eukaryotic cells have microtubules. These are hollow rods constructed of columns of a protein called _____. *(1 mark)*

....

Q4: In animal cells, microtubules radiate out from a region near the nucleus called the _____. *(1 mark)*

The following diagram illustrates the four phases of the cell cycle as arrows. One of the cell cycle events, cytokinesis, is indicated.

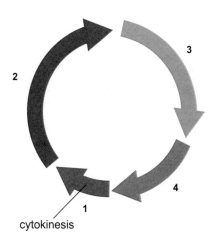

Q5: Which of the following correctly lists the phases of the cell cycle in order? *(1 mark)*

a) G1, S, G2, M
b) M, G1, S, G2
c) M, G2, S, G1
d) G1, M, G2, S

....

Q6: Identify the stage of mitosis illustrated by the following diagram. *(1 mark)*

a) Anaphase
b) Metaphase
c) Prophase
d) Telophase

Q7: Identify the stage of mitosis illustrated by the following diagram. *(1 mark)*

a) Anaphase
b) Metaphase
c) Prophase
d) Telophase

Q8: Identify the stage of mitosis illustrated by the following diagram. *(1 mark)*

a) Anaphase

b) Metaphase
c) Prophase
d) Telophase

Q9: Identify the stage of mitosis illustrated by the following diagram. *(1 mark)*

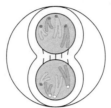

a) Anaphase
b) Metaphase
c) Prophase
d) Telophase

Q10: A cell that passes the checkpoint towards the end of G1 will probably: *(1 mark)*

a) have just completed cytokinesis.
b) move directly into G2.
c) move directly into M phase.
d) replicate its DNA.

Q11: Describe the role of cyclin dependent kinases. *(1 mark)*

Q12: Name the protein that is activated by DNA damage which can stimulate DNA repair. *(1 mark)*

Q13: What is the collective term for proteinases activated during apoptosis? *(1 mark)*

Q14: Give an example of a cell death signal which originates from within the cell. *(1 mark)*

Q15: Which of the following describes the action of oncogenes?

a) They encode the proteins that limit cell division.
b) They restrict cell division at checkpoints.
c) They cause an uncontrolled reduction in the rate of the cell cycle.
d) They cause cell proliferation resulting in tumour formation.

Unit 1 Topic 6

Cells and proteins test

Cells and proteins test

Go online

Laboratory techniques for biologists

Q1: What term describes the process whereby scientists identify control measures to minimise risk from hazards? *(1 mark)*

....

Q2: What piece of equipment can be used to quantify the concentration of a pigmented compound? *(1 mark)*

....

Q3: Use the following standard curve to estimate the protein concentration of a sample with an absorbance of 0.75. *(1 mark)*

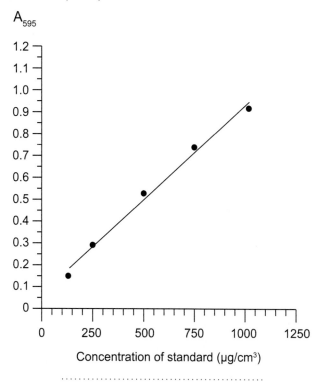

....

TOPIC 6. CELLS AND PROTEINS TEST

Q4: What term describes the pH at which a soluble protein has no net charge and will precipitate out of solution? *(1 mark)*

....

Q5: Outline the two main steps of the SDS-PAGE process which allows proteins to be separated by size alone. *(2 marks)*

....

Q6: Antibodies can be used to detect the presence of a particular protein in a sample. How is binding between the antibody and target protein detected? *(1 mark)*

....

Q7: What is a haemocytometer used for? *(1 mark)*

....

Q8: What substance, containing growth factors, must be added to culture medium for successful growth of animal cells? *(1 mark)*

....

Q9: Why are aseptic techniques used when performing cell culture? *(1 mark)*

Proteomics, proteins structure, binding and conformational change

Q10: What name is given to the entire set of proteins expressed by a genome? *(1 mark)*

....

Q11: Name the class of enzyme found within lysosomes. *(1 mark)*

....

Q12: Describe the role of vesicles. *(1 mark)*

....

Q13: Give the location of lipid synthesis within the cell. *(1 mark)*

....

Q14: What type of bonds link the amino acids together in a protein? *(1 mark)*

....

Q15: What type of bonding is involved the secondary structure of a protein? *(1 mark)*

....

Q16: What aspects of secondary structure of proteins can be seen in the following diagram of myoglobin? *(1 mark)*

....

Q17: The diagram shows that myoglobin contains a non-protein unit which is necessary for its function. What term describes this non-protein unit? *(1 mark)*

....

TOPIC 6. CELLS AND PROTEINS TEST

Q18: What name is given to the process whereby changes in binding at one subunit of a protein alter the affinity of the remaining subunits? *(1 mark)*

....

Q19: Describe the role of protein kinases. *(1 mark)*

Membrane proteins

Q20: Strong hydrophobic interactions hold _____ proteins within the phospholipid bilayer? *(1 mark)*

....

Q21: Define the term facilitated diffusion. *(1 mark)*

....

Q22: What type of channel is shown in the following diagram? *(1 mark)*

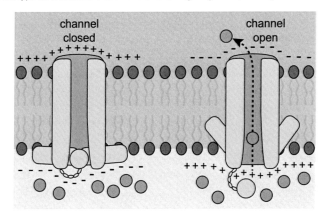

....

Q23: Conformational change of membrane proteins in active transport requires energy from hydrolysis of _____. *(1 mark)*

....

Q24: Which of the following correctly describes the action of the sodium-potassium pump? *(1 mark)*

a) Two sodium ions out of the cell, three potassium ions into the cell.
b) Three sodium ions out of the cell, two potassium ions into the cell.
c) Two sodium ions into the cell, three potassium ions out of the cell.
d) Three sodium ions into the cell, two potassium ions out of the cell.

....

Q25: Describe the relationship between the sodium-potassium pump and glucose symport in the small intestine. *(1 mark)*

Communication and signalling

Q26: What happens to a receptor protein when a signal molecule binds? *(1 mark)*

Q27: Why can the receptors for hydrophobic signalling molecules be within the nucleus? *(1 mark)*

Q28: What type of protein are the receptors for steroid hormones? *(1 mark)*

Q29: Describe the cause of type 2 diabetes. *(1 mark)*

Q30: Explain why exercise may improve glucose uptake in patients with type 2 diabetes. *(1 mark)*

Q31: What is an action potential? *(1 mark)*

Q32: What term describes a change in the membrane potential to a less negative value inside? *(1 mark)*

Q33: Describe the effect of a neurotransmitter binding to its receptor. *(1 mark)*

Q34: Name the area within the eye that detects light. *(1 mark)*

Q35: Name the light sensitive molecule found within rhodopsin. *(1 mark)*

Q36: Select the correct options from each bracket. *(1 mark)*
A single photoexcited rhodopsin activates (hundreds/thousands/millions) of molecules of G-protein. Each activated G-protein activates one molecule of PDE. Each active PDE molecule breaks down (hundreds/thousands/millions) of cGMP molecules per second.

Q37: Which statement correctly describes features of cone cells in humans? *(1 mark)*

a) Cone cells do not function in low light intensity and contain different forms of opsins.
b) Cone cells function in low light intensity and contain different forms of opsins.
c) Cone cells do not function in low light intensity and do not contain different forms of opsins.
d) Cone cells function in low light intensity and do not contain different forms of opsins.

Protein control of cell division

The following diagram represents the phases of the cell cycle.

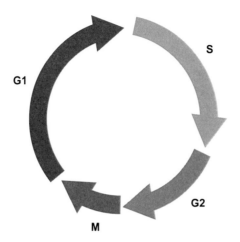

Q38: During which phase of the cell cycle is the DNA replicated? *(1 mark)*

....

Q39: It is important that the rate of the cell cycle is carefully controlled. What type of disease may occur if there is a decrease in the rate? *(1 mark)*

....

Q40: Which stage of mitosis is described as follows? *(1 mark)*

Chromosomes are aligned at the metaphase plate (equator of the spindle).

a) Anaphase
b) Metaphase
c) Prophase
d) Telophase

....

Q41: Which stage of mitosis is described as follows? *(1 mark)*

The chromosomes decondense and nuclear membranes are formed around them.

a) Anaphase
b) Metaphase
c) Prophase
d) Telophase

....

Q42: Which stage of mitosis is described as follows? *(1 mark)*

DNA condenses into chromosomes each consisting of two sister chromatids. Nuclear membrane breaks down; spindle microtubules extend from the MTOC by polymerisation and attach to chromosomes via their kinetochores in the centromere region.

a) Anaphase
b) Metaphase
c) Prophase
d) Telophase

Q43: Which stage of mitosis is described as follows? *(1 mark)*

As spindle microtubules shorten by depolymerisation, sister chromatids are separated, and the chromosomes are pulled to opposite poles.

a) Anaphase
b) Metaphase
c) Prophase
d) Telophase

Q44: During the cell cycle, sufficient phosphorylation by G1 cyclin-dependent kinases allows: *(1 mark)*

a) DNA replication to occur.
b) the cell to arrest the cell cycle.
c) the cell to enter the mitotic phase.
d) the cell to repair DNA damage.

Q45: Microtubules play an important role in mitosis. Which structures, required for mitosis, are composed of microtubules? *(1 mark)*

Q46: Describe the role of CDKs. *(1 mark)*

Q47: Describe one role of p53 protein. *(1 mark)*

Q48: Give an example of an internal death signal. *(1 mark)*

Problem solving

A compound known as arsenic trioxide (ATO) has proved to be an effective treatment for some forms of leukaemia, but has little impact in treating solid tumours such as hepatocellular carcinoma (HCC), a form of liver cancer. A recent study has attempted to improve the efficacy of ATO in treating solid tumours by combining it with N-(β-elemene-13-yl) tryptophan methyl ester (ETME).

The first study used human HCC cells in culture to assess the effects of ATO and ETME on cellular apoptosis. An initial study was performed which determined 20 μmol/l of ETME and 5 μmol/l of ATO should be used. HCC cells were seeded in a culture flask and exposed to ATO only, ETME only or a combination of both. After 48 hours, the cells were stained and cellular apoptosis was evaluated using a detection kit.

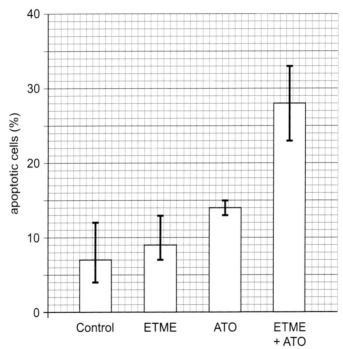

Figure 1: Effect of ETME and ATO on apoptosis of human HCC cells

Western blot analysis was performed to determine the effect of ETME, ATO and the combination on p53 protein expression in human HCC cells. After treatments, the cells were harvested and the protein within was isolated. The proteins were separated by electrophoresis and transferred to a membrane. The membrane was interrogated for the presence of p53 protein using an antibody linked to a reporter enzyme.

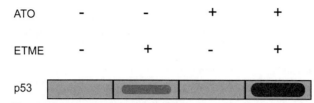

Figure 2: Blot showing the effect of ATO, ETME and the combination on the expression of p53 protein

A final study was conducted to determine the role of p53 in bringing about the apoptotic effect observed when ATO and ETME were used in combination. Human HCC cells were cultured as before with ATO, ETME or the combination. Cell samples from each treatment group were also exposed to a p53 inhibitor and the percentage of apoptotic cells was compared to those without the inhibitor.

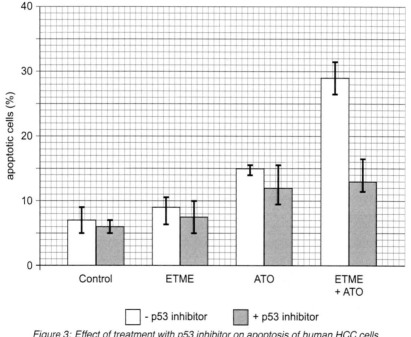

Figure 3: Effect of treatment with p53 inhibitor on apoptosis of human HCC cells

TOPIC 6. CELLS AND PROTEINS TEST

Q49: An initial study determined the concentrations of ATO and ETME which would be used in the main study. What name is given to an experiment which allows the development of protocols before the main study is undertaken? *(1 mark)*

Q50: Explain why the human HCC cells were left exposed to the compounds for 48 hours before the rate of apoptosis was assessed. *(1 mark)*

Q51: Figures 1 and 3 both show error bars. The error bars represent confidence intervals. What do the confidence intervals show about the data collected? *(1 mark)*

Q52: From Figure 1, it can be concluded that ATO and ETME together bring about a significant increase in the rate of apoptosis of human HCC cells in culture when compared to control cells or cells treated with ATO or ETME alone.
Explain how the error bars confirm that this conclusion is valid. *(2 marks)*

Q53: Use data from Figure 1 to show that the combination of ATO and ETME increased the rate of apoptosis by a factor of 4 when compared to the control. *(1 mark)*

Q54: It was concluded that ATO and ETME worked together to bring about apoptosis via a p53 dependent pathway.
How do the results from Figures 2 and 3 support this conclusion? *(2 marks)*

Q55: Further analysis showed that p53 activated a cascade of proteinases to bring about apoptosis. What term describes proteinases which bring about apoptosis? *(1 mark)*

Organisms and Evolution

1 Field techniques for biologists		**145**
1.1	Health and safety	147
1.2	Sampling of wild organisms	147
1.3	Identification and taxonomy	149
1.4	Monitoring populations	149
1.5	Measuring and recording animal behaviour	151
1.6	Learning points	153
1.7	End of topic test	153
2 Evolution		**155**
2.1	Drift and selection	156
2.2	Fitness	158
2.3	Co-evolution	159
2.4	Learning points	161
2.5	End of topic test	162
3 Variation and sexual reproduction		**163**
3.1	Costs and benefits of sexual and asexual reproduction	165
3.2	Meiosis	169
3.3	Sex determination	172
3.4	Learning points	177
3.5	Extended response question	178
3.6	End of topic test	178
4 Sex and behaviour		**181**
4.1	Parental investment	183
4.2	Reproductive behaviours and mating systems in animals	186
4.3	Learning points	192
4.4	Extended response question	193
4.5	End of topic test	194

5 Parasitism — 195

- 5.1 Niche — 198
- 5.2 Parasitic life cycles — 202
- 5.3 Transmission and virulence — 210
- 5.4 Defence against parasitic attack — 211
- 5.5 Immune evasion — 216
- 5.6 Challenges in treatment and control — 217
- 5.7 Learning points — 220
- 5.8 Extended response questions — 222
- 5.9 End of topic test — 223

6 Organisms and evolution test — 229

Unit 2 Topic 1

Field techniques for biologists

Contents

   1.1 Health and safety 147
   1.2 Sampling of wild organisms 147
   1.3 Identification and taxonomy 149
   1.4 Monitoring populations 149
   1.5 Measuring and recording animal behaviour 151
   1.6 Learning points 153
   1.7 End of topic test 153

Prerequisites

You should already know that:

- risk assessments are performed prior to experimentation, outlining safety precautions;

- life can be classified into the three domains: bacteria, archaea and eukaryotes;

- animals can be tracked using banding, ringing, tagging and satellite transmitters;

- ethology is the study of animal behaviour, and that observations are recorded on an ethogram.

Learning objective

By the end of this topic, you should be able to:

- state some hazards and risks associated with field work;
- explain that working in the field may have a greater range of hazards than working in the laboratory;
- describe appropriate methods for sampling wild organisms;
- explain what random, systematic and stratified mean in reference to sampling;
- name methods used in the identification of living things;
- explain that monitoring populations can provide important information for assessing environmental impact;
- describe the method of mark and recapture in estimating population size;
- give examples of effective and ethical methods of marking;
- state how animal behaviour is measured and recorded.

TOPIC 1. FIELD TECHNIQUES FOR BIOLOGISTS

1.1 Health and safety

All practical work, whether in a laboratory or in the field, involves identification of the hazards and risks involved. A risk assessment is a document that identifies the potential hazards, assesses the likelihood of them occurring and clearly describes the steps that can be taken to minimise their occurrence, therefore reducing the possibility of injury or loss. The risk assessment will also outline who is most at risk from the identified hazards. Fieldwork may involve a wider range of hazards compared with working in the laboratory. Additional hazards and risks associated with fieldwork include:

- **terrain** - refers to how the land lies. Variations in terrain may include uneven surfaces, flat areas, hills and steep gradients. Assessing this prior to setting out and selecting appropriate footwear is essential;
- weather conditions - these can change very quickly in the field. A weather forecast should be consulted before setting out, and appropriate clothing, footwear and supplies selected. In extreme weather, fieldwork may have to be postponed or abandoned;
- isolation - areas where fieldwork is carried out can often be isolated. Making sure that others who are not going into the field are aware of the route and the expected time of return is essential;
- tide - this can change very quickly. Tide tables should be consulted prior to setting out.

Control measures from carrying out a risk assessment could include wearing appropriate protective equipment such as clothing and footwear and using specialist equipment and tools.

1.2 Sampling of wild organisms

Sampling should be carried out in a manner that minimises impact on wild species and habitats. Consideration must be given to rare and vulnerable species and habitats which are protected by legislation. Since legislation may change with regional areas, these should be researched prior to beginning sampling.

Sampling techniques

Transect studies

A transect is a line along which different samples can be taken. These are often set up along an area where the terrain or abiotic factors are changeable, e.g. from a woodland into a field or up a sandy shore from the waterline to the high tide line. Abiotic factors may be sampled, as can plant abundance and the abundance of sessile, or very slow moving, organisms. Quadrats of a suitable size and shape for the area are placed along the transect, allowing organism abundance to be recorded. Meters are then used to measure the relevant abiotic factors, e.g. moisture, light, pH and salinity.

Point count

Point count involves an observer recording all individuals seen from a fixed point count location. An example could be bird populations in a given area over a set period of time. Counts are carried out by recording all birds seen and heard from a stationary point. Comparisons can be made throughout the year.

Remote detection

Remote detection employs the use of camera traps that are triggered when wildlife is present. This means that species that are more elusive, i.e. those that are difficult to find, can be observed without the observer being present.

Another good technique for elusive species is scat sampling, which is an indirect method whereby animal droppings are collected in particular areas, providing information about species abundance and diet.

There are different sampling techniques. These include random, stratified and systematic. It is important that the scientist selects the most appropriate method for the area and species being sampled.

- **Random sampling** - individuals selected from the larger population must be chosen completely by chance.
- **Stratified sampling** - in some cases, one large population may be divided up into smaller sub-populations first. Individuals are then randomly selected from each sub-population.
- **Systematic sampling** - may be taken at regular intervals, e.g. every 2 metres along the transect.

Sampling of wild organisms: Questions   Go online

A quadrat 50 cm long was used to sample the number of seaweed plants at five places along two different **line transects** on the rocky shore at South Queensferry. Quadrats were placed every 3 m along each. The results are shown in the following table.

Quadrat number	Number of seaweed plants	
	Area 1	Area 2
1	28	15
2	18	15
3	40	43
4	18	32
5	11	15

Q1: What is the mean density per square metre of Area 1?
....

Q2: What is the mean density per square metre of Area 2?
....

Q3: Explain why this is an example of systematic sampling.

1.3 Identification and taxonomy

Identification of a sample can be made using expertise, classification guides, keys or laboratory analysis of DNA (including mitochondrial and chloroplast DNA), protein or other molecules, such as carbohydrates.

Life is generally classified according to relatedness. Being familiar with **taxonomic** groupings allows predictions and inferences to be made between the biology of an unknown or lesser-known organism and better-known (model) organisms. Model organisms are those that scientists already know a lot about and have been studying for many years. Organisms such as *E. Coli*, *drosophila*, yeast, maize, mice and zebra finches are all good examples of model organisms from different taxonomic groups. Sometimes, organisms may appear more or less related than they actually are due to convergent and divergent evolution respectively. Genetic evidence is often used to dispel myths regarding relatedness. It was recently used to show that red pandas are more closely related to racoons rather than the former theory, which suggested greater relatedness to the giant panda.

To summarise, organisms can be classified in two ways.

1. Taxonomy: where organisms are identified and named by classification groups determined by their shared characteristics.

2. Phylogenetics: where the evolutionary history and relationships among groups, or individuals, are studied.

Predictions and interferences can be made about the biology of an organism by taxonomic groupings. Better-known organisms can be used as 'model' organisms. These are easily studied or have been studied many times before (such as *E.Coli*, *Drosophila Melanogaster*, mice, rats and zebrafish).

1.4 Monitoring populations

Monitoring populations is essential in understanding environmental conditions. Presence, absence or abundance of particular indicator species can give information of environmental qualities, such as the presence of pollutants. For example, biodiversity of lichen species in an area can indicate the levels of sulphur dioxide in the air. Likewise, freshwater invertebrates are frequently used to indicate sewage pollution.

Absence, or reduced population, indicates that a species is susceptible to some environmental factor while an abundance, or increased population, indicates it is favoured by the conditions. This is why susceptible and favoured species can be used to monitor an ecosystem.

Mark and recapture is a method for estimating population size. A sample of the population is captured, marked and released (M). The marking technique used must have minimal impact on the species being studied, thus does not interfere with the individual's normal behaviour or make it more conspicuous to predators. It must, however, also be clear in order to permit subsequent observations. After an interval of time, a second sample is captured (C). If some of the individuals in this second sample are recaptures (R), then an estimate of the total population (N) would be calculated using the formula:

$$N = \frac{MC}{R}$$

- N = estimate of total population;
- M = number captured, marked and released in first sample;
- C = number captured in second sample;
- R = number of marked recaptures in second sample.

Several assumptions are made when using this equation:

- all individuals have an equal chance of capture;
- there is no immigration of emigration during the sample time;
- there is no birth and death during the sample time;
- sampling methods used each time are identical.

Methods of marking include banding, tagging, surgical implantation, painting and hair clipping.

Snail marked and released

Monitoring populations: Question

Go online

Q4: In a survey to estimate a monarch butterfly population in Strathclyde Park, the following data were obtained.

- Number of monarch butterflies first captured, marked and released = 540
- Number of marked monarch butterflies in second capture = 60
- Number of unmarked monarch butterflies in second capture = 180

Calculate the estimated population of monarch butterflies in Strathclyde Park.

1.5 Measuring and recording animal behaviour

Ethology is the study of animal behaviour. An **ethogram** is a recording of all of the observed behaviours shown by a species over a particular period of time. Observed behaviours are recorded as descriptions, e.g. eating, head up, lying down, tail wagging, rather than by trying to infer what the intention of the behaviour may be, e.g. displaying happiness.

The following is an example of an ethogram for an African wild dog.

Behaviour	Description
Resting	Lying on the ground / inactive.
Bow	Head lowered, sometimes front legs lowered too.
Tail wagging	Tail moving from side to side (usually horizontal or higher).
Tail lowered	Tail below horizontal, at times between back legs.
Feeding and drinking	Consuming food or water.
Standing alert	Standing still, looking in direction of movement or noise.
Sniffing ground	Head lowered and nose to the ground in exploration.

Ethogram for an African wild dog

It is essential that descriptions are clear and unambiguous. Extreme care must be taken to avoid **anthropomorphism** (personification), which is when animals are credited with human emotions and qualities.

Ethologists take measurements such as latency (the time taken for the animal to respond to a particular stimulus), frequency and duration of certain behaviours and activities. A time budget can be constructed using this data, which is the percentage of time that the animal spends on certain activities.

Sometimes, a time budget may be presented as a pie chart showing the duration of different behaviours observed by an animal over a set period.

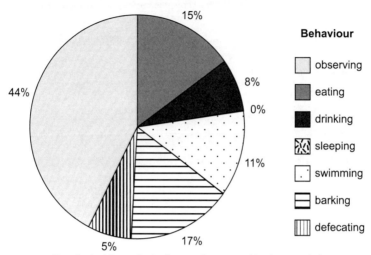

Time budget for an Australian sea lion over a 30 minute period

Alternatively, a table or chart showing how frequently certain behaviours are observed or how long they last can also be produced. Additionally, how long an animal takes to respond to a certain stimulus may be important, and therefore included in a chart. Ethograms and time budgets are very much specific to the aims of the study and the animals being observed.

Measuring and recording animal behaviour: Question   Go online

Q5: A student who was observing the behaviour of a lioness with one of her offspring made some field notes. Which of the following notes demonstrate anthropomorphism?

A) Offspring displayed teeth to mother.

B) Offspring begged mother for food.

C) Offspring pounced at mother.

D) Offspring smiled at mother.

1.6 Learning points

Summary

- Often there are a wider range of hazards associated with fieldwork than laboratory work.
- Hazards, such as those resulting from uneven/challenging terrain, weather conditions and isolation, should be identified and the risks assessed.
- When sampling wild organisms, the technique chosen should be appropriate to the species being sampled.
- Techniques for sampling wild organisms include transects, point counts, remote detection, quadrats, camera traps and scat sampling.
- Sampling should be random, stratified and systematic.
- Organisms can be identified using keys, classification guides or analysis of DNA or proteins.
- Monitoring populations is important in assessing environmental impact.
- Mark and recapture is a suitable technique for estimating population size: $N = MC/R$.
- Organisms can be marked using appropriate and effective techniques, including tagging, banding, surgical implantation, painting and hair clipping.
- Assumptions are made when carrying out mark and recapture studies.
- Ethograms and time sampling are used to compare the behaviour of different individuals of a species.
- In observing and recording animal behaviour anthropomorphism must be avoided.

1.7 End of topic test

End of Topic 1 test — Go online

Q6: State *two* additional risks or hazards associated with fieldwork compared to laboratory work. *(2 marks)*

....

Q7: Elusive species are difficult to sample. Name one technique used for sampling elusive species. *(1 mark)*

....

Q8: Which of the following statements correctly describes an example of stratified sampling? *(1 mark)*

a) A quadrat is placed every 2 m along a 20 m transect line.
b) 50 ladybirds are randomly chosen from a population of 500 ladybirds and their metabolic rate is measured.
c) 10 pupils from each year group in a school are randomly selected and their pulse rate is taken.

....

Q9: Mark and recapture is a method used in monitoring populations. Give *two* methods for marking wild organisms. *(2 marks)*

....

Q10: The formula

$$N = \frac{MC}{R}$$

is used to estimate population size using mark and recapture data where:

- N = population estimate;
- M = number first captured, marked and released;
- C = total number in second capture;
- R = number marked in second capture.

In a survey to estimate a peppered moth population in Prince's Street Gardens, the following data were obtained.

- Number of peppered moths first captured, marked and released = 200
- Number of marked peppered moths in second capture = 70
- Number of unmarked peppered moths in second capture = 105

Calculate the estimated population of peppered moths in Prince's Street Gardens. *(1 mark)*

....

Q11: What is the study of animal behaviour called? *(1 mark)*

....

Q12: What are observations of animal behaviour recorded on? *(1 mark)*

....

Q13: What term is used to describe the personification of animal behaviour? *(1 mark)*

a) Anthropomorphism
b) Endomorphism
c) Homeomorphism
d) Pseudomorphism

Unit 2 Topic 2

Evolution

Contents

2.1 Drift and selection 156
2.2 Fitness 158
2.3 Co-evolution 159
2.4 Learning points 161
2.5 End of topic test 162

Prerequisites

You should already know that:

- mutations are rare, random changes to genetic sequences which can be harmful, beneficial or neutral;
- genetic material can be passed horizontally in prokaryotes (bacteria);
- natural selection is a non-random process, resulting in a change to the frequency of certain genes occurring within a population;
- genetic drift is a random process, having greater effect on smaller populations, resulting in a change to the frequency of certain genes occurring within a population;
- symbiosis is the relationship between two organisms, each of a different species, whereby at least one of the two benefits.

Learning objective

By the end of this topic, you should be able to:

- describe and explain the process of evolution with reference to sexual selection, natural selection and genetic drift;
- state that mutations give rise to genetic variation;
- define the terms absolute fitness and relative fitness;
- explain, using examples, co-evolution and the Red Queen hypothesis.

156 UNIT 2. ORGANISMS AND EVOLUTION

2.1 Drift and selection

Evolution is the change, over successive generations, in the proportion of individuals in a population differing in one or more inherited traits. Evolution can occur through the random process of **genetic drift** or the non-random processes of **natural selection** and **sexual selection**.

Genetic drift is the random change in how frequently a particular allele occurs within a population. Genetic drift has a more significant impact in small populations because alleles are more likely to be lost from the gene pool.

In the small population of just seventeen individuals illustrated below, if some random chance event killed eight individuals including all four dark blue individuals, the population would become uniformly light red. This is an example of genetic drift.

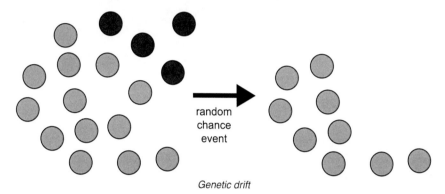

Genetic drift

In comparison to genetic drift, natural and sexual selection are non-random processes whereby certain alleles occur more frequently within a population because they confer a selective advantage. These alleles increase the chance that the individuals can compete and survive to pass the advantageous allele on to future generations. Selection results in the non-random increase in the frequency of allles that are advantageous as well as the decrease in deleterious gene frequency.

Populations, in an environment that can support individuals with variations, can produce more offspring better suited to that environment who will then go on to survive, breed and pass on their alleles to new generations.

The main source of novel alleles arising within a population is as a result of random **mutation**. Mutations are the source of new DNA sequences, which can also mean novel alleles too. Where the majority of mutations are deleterious or neutral, in very rare cases they may be beneficial to the fitness of an individual.

Genetic drift can alter a gene pool because certain alleles may be under represented, or even over represented, causing the allele frequency to change. When selection pressures are stronger, the rate of evolution can be rapid.

The Hardy-Weinberg (HW) principle states that, in the absence of evolutionary influences, allele and gene frequencies in a population will remain constant over time.

TOPIC 2. EVOLUTION

The equation used for this is: $p^2 + 2pq + q^2 = 1$

- p = frequency of dominant allele
- q = frequency of recessive allele
- p^2 = frequency of homozygous dominant genotype
- $2pq$ = frequency of heterozygous genotype
- q^2 = frequency of homozygous recessive genotype

Sexual selection is the non-random process tat involves the selection of alleles to increase the individuals chances of mating and producing offspring. This could lead to sexual dimorphism.

Examples of sexual selection can be due to female choice (females assessing male fitness) or male-male rivalry (large size and weaponry can increase a males access to females).

Drift and selection: Questions — Go online

Decide whether the following four attributes belong to genetic drift or natural/sexual selection.

Q1: Affects smaller populations.

a) Genetic drift
b) Natural or sexual selection

....

Q2: Alleles that give a selective advantage increase in frequency.

a) Genetic drift
b) Natural or sexual selection

....

Q3: Non-random.

a) Genetic drift
b) Natural or sexual selection

....

Q4: Random.

a) Genetic drift
b) Natural or sexual selection

2.2 Fitness

Fitness is an indication of an individuals ability to be successful at surviving and reproducing. It refers to all the contributions made to a gene pool of the next generation by individual genotypes. Fitness can be measured in absolute or relative terms.

- **Absolute fitness** is the ratio of frequencies of a particular genotype from one generation to the next. That is the ratio between the number of individuals with a particular genotype after selection, compared to the number with that same particular genotype before selection. If the absolute fitness is 1, then the frequency of that genotype is stable. A value greater than 1 conveys an increase in the genotype and, therefore, a value less than 1 conveys a decrease.

$$\text{absolute fitness} = \frac{\text{frequency of a particular genotype after selection}}{\text{frequency of a particular genotype before selection}}$$

- **Relative fitness** is the ratio of surviving offspring of one genotype compared with other genotypes. For example, on average, pea plants with purple flowers produce more offspring than those with white flowers. Purple pea plants, as the most reproductively successful strain, are given a relative fitness of 1. If the white flowering pea plants only produce 65% as many offspring, their relative fitness would be 0.65 by comparison. As a result, the purple pea is the favoured trait and this allele would become more frequent in subsequent generations.

$$\text{relative fitness} = \frac{\text{number of surviving offspring per individual of a particular genotype}}{\text{number of surviving offspring per individual of the most successful genotype}}$$

Calculating relative fitness

Relative fitness can be a measure of the reproductive success of a particular organism compared to other members of the population, or the success of a particular genotype within a population. Either way, this is calculated by dividing the absolute fitness by the average fitness within the population.

For example, if a dormouse has 6 surviving offspring in a population where the average number of surviving offspring is only 4, this mouse has a relative fitness of 6 ÷ 4 = 1.5. Alternatively, a dormouse in this population with only 2 surviving offspring will have a relative fitness of just 0.5.

The concept of relative fitness can be extended further to consider genotype. The genotype with the greatest fitness is given a value of 1. Let's say that dark green (DD) toads are the fittest, averaging 10 surviving offspring each. To calculate the relative fitness of brown (dd) toads with an average of 6 surviving offspring, 6 ÷ 10 = 0.6.

Fitness: Questions   Go online

Q5: The ratio of frequencies of a particular genotype from one generation to the next is defined as _____ fitness.

....

Q6: The ratio of surviving offspring of one genotype compared with other genotypes is defined as _____ fitness.

2.3 Co-evolution

Co-evolution is frequently seen in pairs of species that demonstrate a symbiotic relationship. **Symbiosis** is co-evolved intimate relationships between members of two different species.

Mutualism describes a relationship where both participants gain from the interaction. Mutualistic relationships are described as interdependent, this means one cannot live without the other and the interaction is know as (+/+).

Commensalism results in interactions where only one of the organisms benefit and is written (+/0).

Parasites and their hosts

Parasitism involves a relationship whereby a **parasite** lives on a host, gaining resources and a place to live. This is of benefit to the parasite, while the host is harmed, and is written as (+/-). Co-evolution has been documented between the parasite that causes malaria, the mosquito that acts as a vector for this parasite and humans, which are the host. Current fossil evidence suggests that the parasite started to spread extensively about 100,000 years ago. This coincides with evidence to suggest when the first major human migrations took place. More recently, about 10,000 years ago, DNA sequence evidence suggests that both the parasite and mosquitoes underwent rapid evolution. Both of these events provide strong evidence for the complicated co-evolution between the parasite and the two other species; humans and mosquitoes.

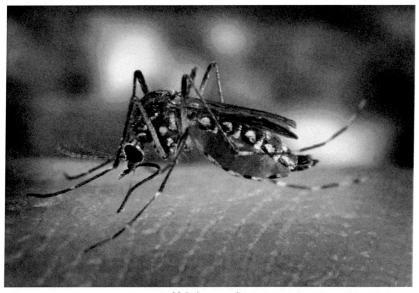

Malaria mosquito

 In co-evolution, a change in the traits of one species acts as a selection pressure on the other species, driving natural selection accordingly. The co-evolutionary 'arms race' between a parasite and host is known as the **Red Queen hypothesis** *because both organisms must 'keep running in order to stay still'.*

UNIT 2. ORGANISMS AND EVOLUTION

The Red Queen analogy comes from the Alice in Wonderland story "Through the Looking Glass". In this story, Alice is seeking out the Queen of Hearts. The faster Alice runs to try and reach the Queen, the faster the Queen seems to go until, eventually, she vanishes. Alice is advised to go back the way she came and bumps right into the Queen. Alice is then lead by the Queen to the top of hill. Again the Queen runs. Alice starts running fast too, but when she and the Queen stop, neither has actually moved. Co-evolution is just like this, where evolution merely permits organisms to maintain their current success.

In a co-evolutionary relationship, changes in the traits of one species can act as a selection pressure on the other species so both species must adapt to avoid extinction.

Co-evolution and the Red Queen: Questions   Go online

The following table compares slug parasitism by nematodes of two different slug populations.

- Slug population 1 - no previous exposure to nematode parasites.
- Slug population 2 - previous exposure to nematode parasites.

Time (weeks)	Number of nematodes in slug population 1	Number of nematodes in slug population 2
0	10	10
10	200	20
20	425	15
30	450	25
40	450	20

Q7: Calculate the percentage change in nematode numbers in slug population 1 from 10 to 20 weeks.

....

Q8: What evidence from the table suggests that slug population 2 has developed resistance to the nematodes?

....

Q9: How would this resistance have evolved?

....

Q10: In terms of the Red Queen hypothesis, predict what would happen to the numbers of nematode parasites in slug population 2 if left for a few generations.

2.4 Learning points

Summary

- Evolution is the change, over successive generations, in the proportion of individuals in a population differing in one or more inherited characteristics.

- Genetic drift is the random change in the frequency of a particular allele within a small population.

- Natural and sexual selection are non-random processes whereby certain alleles become more frequent within a population because they confer a selective advantage.

- The main source of novel alleles arising within a population is as a result of random mutations, most of which are deleterious or neutral, and, in very rare cases, may be beneficial to the fitness of an individual.

- Absolute fitness is the ratio of frequencies of a particular genotype from one generation to the next.

- Relative fitness is the ratio of surviving offspring of one genotype compared with other genotypes.

- Where selection pressures are high, the rate of evolution can be rapid.

- Co-evolution will occur in pairs of species that interact with one another on a regular basis.

- Mutualism, commensalism and parasitism are types of symbiotic relationships.

- The co-evolutionary 'arms race' between a parasite and host is known as the Red Queen hypothesis, whereby a change in the traits of one species acts as a selection pressure on the other species.

2.5 End of topic test

End of Topic 2 test — Go online

Q11: Which random process that contributes to evolution affects smaller populations much more than larger populations? *(1 mark)*

....

Q12: What is the main source of new DNA sequences within a population? *(1 mark)*

....

Q13: When comparing the number of red genotypes to blue genotypes occurring within a butterfly population, what type of fitness is being measured? *(1 mark)*

....

Q14: What name is given to the hypothesis used to describe the continuous co-evolution of parasites and their hosts? *(1 mark)*

Unit 2 Topic 3

Variation and sexual reproduction

Contents

3.1 Costs and benefits of sexual and asexual reproduction 165
3.2 Meiosis 169
3.3 Sex determination 172
3.4 Learning points 177
3.5 Extended response question 178
3.6 End of topic test 178

Prerequisites

You should already know that:

- sexual reproduction involves fusion of gametes from two different parents, resulting in dissimilar offspring, whereas asexual reproduction involves making exact copies of one parent;

- it is advantageous to have genetically varied offspring so that some may survive should conditions change;

- diploid refers to a cell with two full sets of chromosomes and haploid refers to a cell with just one full set of chromosomes;

- nuclear division is referred to as mitosis and results in all of the daughter cells being genetically identical to the mother cell;

- carriers are individuals who have a recessive allele that is masked by a dominant allele, yet they can pass this gene on to offspring;

- sex chromosomes (XY) determine the sex of an individual in many insect and mammal species.

UNIT 2. ORGANISMS AND EVOLUTION

Learning objective

By the end of this topic, you should be able to:

- state the disadvantages of sexual reproduction;
- explain why, despite the disadvantages, sexual reproduction is so common;
- name successful asexual reproduction strategies;
- describe the stages of meiosis;
- describe the features of homologous chromosomes;
- state that meiosis results in variable gametes and describe the contribution of independent assortment and crossing over to this variation;
- define the term linked genes and explain how they are used to map chromosomes;
- state that environmental factors, for example temperature and resource availability, can affect sex ratio;
- define the term hermaphrodite;
- explain that the sex of some organisms can change due to competition or parasitic infection;
- explain why males are more likely to be affected by sex-linked conditions than females;
- explain that the random deactivation of genes on one of the X chromosomes in each cell prevents carrier females being affected by sex-linked genetic conditions.

3.1 Costs and benefits of sexual and asexual reproduction

Compared to asexual reproduction, sexual reproduction appears to have two disadvantages.

1. Half of the population (males) are unable to produce offspring.
2. Each parent is only able to pass on half of their genetic material rather than the full 100%.

The benefits to sexual reproduction must, however, be greater than the disadvantages because it is so common. The benefits lie in the greater genetic variation within sexually reproducing organisms. This genetic variation provides some organisms with the ability to adapt to changing conditions or survive new strains of diseases. Without such genetic variety, the Red Queen's arms race (see Topic 2: "Evolution") would stop. In other words, this variation can drive selection and evolution should the different genes and alleles provide the individual with an advantage and, therefore, an increased chance of survival.

If hosts reproduce sexually, the genetic variability in their offspring decreases the chances that all of the individuals will be susceptible to infection by parasites. This means that the host will be able to resist and tolerate parasitism, therefore showing a greater fitness.

Where niches are narrow and stable, asexual reproduction can be a successful reproductive strategy. Asexual reproduction is also advantageous when recolonising disturbed habitats because it is usually faster. In eukaryotes, examples of asexual reproduction include:

- **vegetative propagation/cloning** in plants, such as reproduction via bulbs, e.g. onions and daffodils, and runners, e.g. spider plants and strawberry plants - all of the resulting offspring will be genetically identical to the original parent plant;

Daffodil bulbs

Flowering daffodils

Strawberry plant

Strawberry plant with runners by https://www.flickr.com/photos/gabrielahpaulin/, licenced under the Creative Commons
https://creativecommons.org/licenses/by/2.0/deed.en license

- **parthenogenic** animals lack fertilisation - embryos result from unfertilised eggs and, therefore, the resulting offspring will be haploid. When kept in captivity for several years with no male contact, female Komodo dragons have been known to reproduce without fertilisation. It appears that, for continuation of the species, this is how the female responds to isolation. This would be advantageous should females become isolated in the wild or should males die. The offspring of parthenogenesis in Komodo dragons are always male. Conversely, stick insects can also reproduce asexually in the absence of males; however, all of the offspring are female in this case. Parthenogenesis is found to be more common in cooler climates with low parasite diversity.

Komodo dragon

Stick insect

In organisms where asexual reproduction is most common, e.g. bacteria and yeast, many have mechanisms for horizontal gene transfer between individuals. This means that one individual can pass genetic information to another within the population, often via a connecting tube. Bacteria commonly pass plasmids between one another.

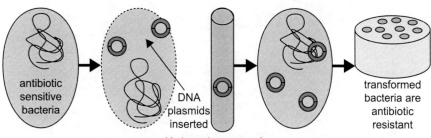

Horizontal gene transfer

Costs and benefits of reproduction: Questions

Go online

Q1: In fire ants, males may be produced by parthenogenesis from an egg cell. The diploid number of chromosomes is 8. When parthenogenesis occurs, how many chromosomes does a male fire ant have?

a) 2
b) 3
c) 4
d) 8

....

Q2: Which term would be defined as "asexual reproduction in plants such as bulbs or runners"?

a) Horizontal gene transfer
b) Parthenogenesis
c) Vegetative propagation

....

Q3: Which term would be defined as "asexual reproduction in reptiles where offspring are produced in the absence of fertilisation"?

a) Horizontal gene transfer
b) Parthenogenesis
c) Vegetative propagation

....

Q4: Which term would be defined as "common in prokaryotes where sections of DNA, often in the form of plasmids, are passed between one another"?

a) Horizontal gene transfer
b) Parthenogenesis
c) Vegetative propagation

3.2 Meiosis

Meiosis is the process where gametes are produced in the reproductive organs. In animals, this takes place in the ovaries and testes. In plants, this takes place in the anthers and ovaries. One diploid gametocyte (gamete mother cell) divides into four haploid sex cells.

Meiosis has two divisions and results in variable gametes. Division one is called meiosis I and division two is called meiosis II.

Meiosis I

1. Division 1 (meiosis I) starts with **interphase**. This is where each chromosome undergoes DNA replication to become two identical chromatids.

2. The **homologous chromosomes** pair up and line up along the equator of the cell.

3. Homologous chromosomes are pairs of chromosomes:

    - of the same size;
    - with the same centromere position;
    - with the same genes at the same loci.

4. The alleles of the genes on homologous chromosomes may, however, be dissimilar because each homologous chromosome is inherited from a different parent.

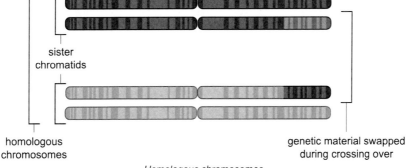

Homologous chromosomes

5. When the homologous chromosomes pair up, they can touch each other at points called **chiasmata**. **Crossing over** can then occur at chiasmata. This is when sections of DNA get swapped, shuffling sections of DNA between the homologous pairs, allowing the recombination (creation of new combinations) of alleles to occur.

6. Another process in meiosis I that results in variation between gametes is **independent assortment**. When homologous chromosomes pair up and line up along the equator, the final position of one pair is completely random relative to every other pair. Additionally, these chromosomes go on to be separated, irrespective of their maternal and paternal origin. This means that independent assortment results in gametes with varying combinations of maternal and paternal chromosomes.

7. The number of possible combinations is worked out by calculating 2 (since there are two of each chromosome inherited - one from each parent) to the power of however many chromosome pairs there are. In humans, there are 23 pairs of chromosomes, thus the haploid number is 23. Therefore, the number of possible combinations in the gamete is:

$$2^{23} = 8,388,608 \; combinations$$

8. Homologous chromosomes are separated by spindle fibres and are pulled to opposite ends of the cell.
9. The cell divides, forming two haploid daughter cells.

Meiosis II

1. In division 2 (meiosis II), chromatids are separated by spindle fibres and four haploid gametes (often genetically dissimilar) are formed.

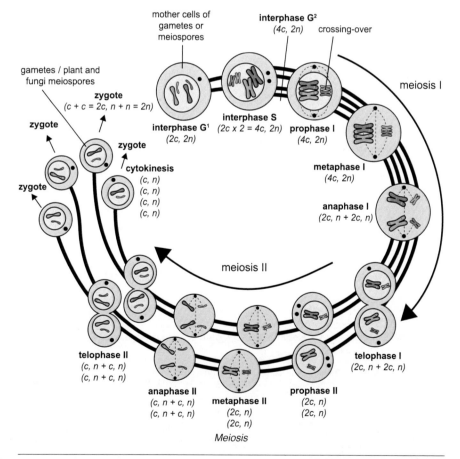

Meiosis

Linked genes and chromosome mapping

Genes on the same chromosome are said to be linked. Genes that are situated closely on a chromosome are less likely to be separated during crossing over, whereas genes that are further apart are much more likely to be separated. Separating **linked genes** through crossing over creates **recombinants**. Scientists use the frequency of recombination to map chromosomes, working out where each gene is in relation to others and pinpointing exact locations on the chromosome. An example of how recombination frequency is used to map chromosomes is given below.

In certain organisms, genes A, B, C and D are located on the same chromosome. The percentage recombination between pairs of these genes is shown in the following table.

Gene pair	Percentage recombination
A and B	15
A and C	5
B and D	5
C and D	25

Since genes C and D have a percentage recombination of 25%, they will be quite far apart on the chromosome. Conversely, with a percentage recombination of just 5%, genes A and C will be closer to one another. The best thing to do is to draw a straight line to represent your chromosome, then start to write letters in, making sure that the numbers work. This can take a little trial and error so use a pencil.

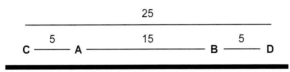

Mapping a chromosome

Meiosis forms variable gametes: Activity

Q5: Arrange the following key steps in the process of meiosis in the correct order.

- Chromatids are separated by spindle fibres.
- Chromosomes undergo DNA replication (interphase).
- Crossing over occurs at points called chiasmata.
- Four haploid gametes are produced.
- Homologous chromosomes line up at the equator of the cell.
- Homologous chromosomes touch at points called chiasmata.
- Independent assortment occurs.
- Two haploid cells are formed.

Meiosis forms variable gametes: Questions

A mosquito has six chromosomes.

Q6: What is its diploid number?
....

Q7: What is its haploid number?
....

Q8: How many different combinations are possible in the gametes of a mosquito?
....

Q9: In *Drosophila*, the genes for wing length (L), eye colour (C), body colour (B) and the presence of bristles (P) are linked (all on the same chromosome). The following table gives the frequency of recombination obtained in crosses involving different pairs of linked genes.

Gene pair in the cross	Frequency of recombination (%)
Wing length x Eye colour	12
Wing length x Body colour	18
Wing length x Presence of bristles	16
Eye colour x Presence of bristles	4
Body colour x Presence of bristles	2

Use the information to show the positions of these genes in relation to each other on a chromosome diagram. Work out in which order the linked genes L, C, B and P would appear on the chromosome.

3.3 Sex determination

Many organisms, usually invertebrates, are **hermaphroditic**. Hermaphrodites are species that have functioning male and female reproductive organs in each individual. They produce both male and female gametes and usually have a partner to exchange gametes with.

Simultaneous hermaphrodites are organisms with both male and female reproductive organs, e.g. earthworms and slugs. Sequential hermaphrodites are born as one sex and may change to the other sex at some point during their life, e.g. some fish and jellyfish. This usually happens if reproductive success is likely to be much greater by being the other sex.

Sometimes, sex determination is under environmental rather than genetic control. Crocodiles and alligators are a good example of this. In Nile crocodiles, if the temperature inside the nest, and therefore the egg incubation temperature, is below 31.7°C or above 34.5°C, the offspring will be female. Males are only born in the narrow range between these two temperatures.

Nile crocodile

In some species, the sex ratio of offspring can be adjusted in response to population density and the resulting strain on resource availability. This has been observed in deer. There has been a correlation between dominant females producing greater numbers of male offspring. How this manipulation occurs is yet to be explained, however, the female seems to be able to control sex selection. Some theories suggest that the female can detect the differently shaped sperm responsible for sex determination and can control them as they move through the reproductive tract. During times of higher population density and excessive winter rainfall, this relationship between dominant females producing more males was no longer found. This finding has been linked to poorer nutrition in pregnant females.

Changing sex

Some species of animals have been found to change their sex during their lifetime. Sex can change as a result of size, competition or parasitic infection. However, the underlying reason for a change in sex will be to permit the occurrence of successful reproduction. Male shrimps have been found to change into females in response to parasitic infection. In anemone fish, all juveniles are males. On forming monogamous pairs, one of the fish will become female. Should the female die, the male in the pair becomes female and a new juvenile male moves into the pair. In another type of fish, the bluebanded goby, when a dominant male dies, a large female will take his place, changing gender to do so. The female starts to display more aggressive behaviour and undergoes changes in hormone levels.

Sex chromosomes and sex linkage

In mammals, and some insects such as *Drosophila* and butterflies, sex is determined through a set of specific chromosomes - the sex chromosomes. Usually, these are known as the X and Y chromosomes, with a gene on the Y chromosome often determining the development of 'maleness'.

Males are said to be **heterogametic** because their sex chromosomes are dissimilar (XY). The male lacks homologous alleles on the smaller (Y) chromosome. This can result in sex-linked patterns of inheritance whereby males have a greater chance of being affected by certain recessive conditions, e.g. colour blindness and haemophilia in humans. This is because females would need two copies of the recessive allele to be affected, whereas males only need one.

Females may be carriers, thus have one copy of the affected gene, yet remain unaffected. They can, however, pass this onto any offspring they have. In females, the parts of the X chromosome that are absent from the Y chromosome are randomly inactivated in one of the homologous X chromosomes in each cell in the body. This prevents a double-dose of gene products. This also means that carriers will not suffer from the effects of any harmful mutations on these X chromosomes because the X chromosome inactivation is random and 50% of the cells in any tissue will have a normal copy of the gene in question.

A good example of a sex-linked gene in humans is haemophilia. This is a condition where the affected individual does not produce the blood clotting agent Factor VIII. Haemophilia is the result of a recessive gene carried on the X chromosome. If we use the letter "N" for normal and "n" for affected, we can produce the following genotypes:

- Normal male = $X^N Y$
- Affected male = $X^n Y$
- Normal female = $X^N X^N$
- Carrier female = $X^N X^n$
- Affected female = $X^n X^n$

In the following cross, a normal male is crossed with a carrier female.

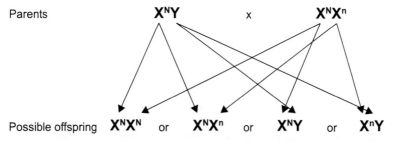

As can be seen here, a range of phenotypes may be expected, yet the only affected individual is a son. To be affected, the male only requires one recessive gene; however, the female needs two haemophilia genes to be affected. In the carrier female, the haemophilia gene is masked by the normal gene; however, she has a 50% chance of passing this gene to her offspring and any sons she passes it to will have haemophilia.

Tortoiseshell cat

In cats, a tortoiseshell coat results from a mixture of black and red. The red coat colour gene is dominant (O) to cream (o) and is carried on the X chromosome. Red is also dominant to black; however, black is dominant to cream. The black coat colour gene is not sex-linked, therefore it must be on an **autosome**. In males, which have only one X chromosome, if their X chromosome has a red coat colour gene, then they will have a red coat. In female carriers, $X^O X^o$ will have a patchy coat, with red in places and black being able to show through in other places. This is due to the random inactivation of one of the two X chromosomes in every cell in the cat.

Sex determination: Questions

Eggs from Nile crocodiles were incubated at two different temperatures over five breeding seasons. When the crocodiles hatched, their sex was recorded. The following graph shows the affect that temperature had on gender within the population.

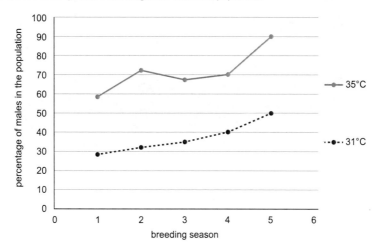

Q10: Calculate the percentage change in females in the population between breeding seasons 4 and 5 when incubated at 35°C.

....

Q11: Calculate the percentage change in males in the population between breeding seasons 4 and 5 when incubated at 31°C.

....

Q12: How many females would be present in a population of 400 Nile crocodiles after 4 breeding seasons at 31°C?

....

Q13: How many males would be present in a population of 600 Nile crocodiles after 5 breeding seasons at 35°C?

....

Q14: Haemophilia is a recessive sex linked condition carried on the X chromosome. In a cross between an affected male and a carrier female, what are the chances that any offspring will be affected?

....

Q15: In the same cross, what are the chances that a son will be affected?

....

Q16: In the same cross, what are the chances that a daughter will be affected?

3.4 Learning points

Summary

- There are two disadvantages of sexual reproduction. Only females can produce offspring and each parent only passes on 50% of their genes rather than the full 100%.

- The one advantage of sexual reproduction is the huge genetic variability amongst offspring, which outweighs any disadvantages.

- Asexual reproduction is ideal in stable niches and includes strategies in eukaryotes, such as vegetative propagation in plants and parthenogenesis in some insects and reptiles

- For prokaryotes, where reproduction tends to be asexual, horizontal gene transfer between individuals is a common mechanism.

- Meiosis is the process whereby one diploid gamete mother cell divides into four genetically dissimilar haploid gametes.

- Independent assortment and crossing over both occur in meiosis I and are responsible for increasing the genetic variation amongst gametes.

- Homologous chromosomes are pairs of chromosomes of the same size, with the same centromere position and with the same genes at the same loci.

- The alleles of the genes on homologous chromosomes may, however, be dissimilar because each homologous chromosome is inherited from a different parent.

- Linked genes are found on the same chromosome. The frequency of linked chromosomes being separated through crossing over can be used to map chromosomes.

- Hermaphrodites are organisms with both male and female reproductive organs.

- In some organisms, such as crocodiles, environmental conditions can affect the sex ratio of offspring.

- In other cases, the sex ratio is related to resource availability.

- Some organisms can change their sex in their lifetime depending on environmental conditions, parasitic infection or competition.

- In mammals, males tend to be heterogametic. This means that sex linked traits are more common in males than females.

- Females can be carriers of a recessive sex-linked trait, yet show no deleterious effects.

- The reason for this is that one X chromosome in each cell is randomly deactivated so that half of the cells will contain an active working copy of the gene.

3.5 Extended response question

The activity which follows presents an extended response question similar to the style that you will encounter in the examination.

You should have a good understanding of meiosis before attempting the question.

You should give your completed answer to your teacher or tutor for marking, or try to mark it yourself using the suggested marking scheme.

Extended response question: Meiosis

Give an account of meiosis. *(10 marks)*

3.6 End of topic test

End of Topic 3 test Go online

Q17: Males being unable to produce offspring is one disadvantage of sexual reproduction. What is the other disadvantage of sexual reproduction that is universal to all species? *(1 mark)*

a) Finding a mate is difficult.
b) Not all individuals are attracted to the opposite sex.
c) Only half of each parent's genome is passed onto offspring.
d) Genetic variation decreases.

....

Q18: Sexual reproduction leads to genetic variation _____. *(1 mark)*

a) decreasing
b) increasing
c) staying the same

....

Q19: Which of the following statements about parthenogenesis is not true? *(1 mark)*

a) Unfertilised eggs develop into embryos.
b) This is common in plants.
c) Offspring are haploid.
d) It is common in cooler climates.

....

TOPIC 3. VARIATION AND SEXUAL REPRODUCTION

Q20: Which of the following statements are true of meiosis I only? *(1 mark)*

A) Crossing over occurs at points called chiasmata.
B) Independent assortment occurs.
C) Homologous chromosomes are separated.
D) Four haploid gametes are produced.

....

Q21: State two features of homologous chromosomes. *(2 marks)*

....

Q22: Name two processes that are responsible for increasing genetic variation during meiosis. *(2 marks)*

....

Q23: Genes on the same chromosome are said to be _____. *(1 mark)*

....

Q24: What term is used to describe an individual with both male and female reproductive organs? *(1 mark)*

....

Q25: Which of these terms means that the sex chromosomes are dissimilar? *(1 mark)*

a) Heterogametic
b) Hermaphroditic
c) Recombination
d) Sex-linked

....

Q26: In sex linked conditions in humans, the male requires _____ of the affected gene to show the condition in their phenotype. *(1 mark)*

a) one copy
b) two copies

....

Q27: In sex linked conditions in humans, the female requires _____ of the affected gene to show the condition in their phenotype. *(1 mark)*

a) one copy
b) two copies

....

Q28: Explain why carrier females remain unaffected by deleterious mutations carried on the X chromosome? *(1 mark)*

Unit 2 Topic 4

Sex and behaviour

Contents

4.1 Parental investment 183
4.2 Reproductive behaviours and mating systems in animals 186
4.3 Learning points 192
4.4 Extended response question 193
4.5 End of topic test 194

Prerequisites

You should already know that:

- sperm are smaller and produced in greater numbers than eggs;

- males often have to compete or display for females, who will then choose the fittest;

- parental care has a high energy demand; however, this technique increases the chance of reproductive success;

- sexual selection is a type of natural selection by which females choose males based on certain physical characteristics.

UNIT 2. ORGANISMS AND EVOLUTION

Learning objective

By the end of this topic, you should be able to:

- compare sexual investment between males and females;
- state that parental investment is costly, but greatly increases the chance of successful reproduction;
- compare the features of r-selected and K-selected organisms;
- state that classification of parental investment into discrete r-selected or K-selected categories fails to reflect the complex range of life history strategies;
- explain the correlation between the reproductive strategy employed, and the number and quality of current offspring versus possible future offspring;
- define the term sexual dimorphism;
- describe male-male rivalry in courtship behaviour, explaining that smaller males may use a variety of techniques such as sneaking;
- explain that successful courtship behaviour in birds and fish can be a result of species-specific sign stimuli and fixed action pattern responses;
- explain that females use courtship behaviour to make choices based on male fitness, usually related to advantageous genes or low parasite infestation;
- using examples, explain what lekking is.

TOPIC 4. SEX AND BEHAVIOUR

4.1 Parental investment

Sperm versus egg production

In animals, sperm are always far more numerous than eggs. Due to the presence of an energy store, eggs are much larger and fewer in number. A far greater investment is made by females due to the production of a lower number of larger gametes. As a result, their chance of reproductive success is lower and the chance of passing on genes is reduced.

Parental investment is costly but increases the chance that young will survive. The level of parental care will depend on the number of offspring produced and the environment. In a stable environment, organisms tend to produce a smaller number of rather 'expensive' young, thus a lot of energy will be invested in caring for each one. Conversely, in unstable environments, many 'cheap' young are produced with low energy investment and parental care.

These particular life history strategies can be categorised into two groups:

1. **r-selected** populations;
2. **K-selected** populations.

Characteristic	r-selected population	K-selected population
Environment	Unstable	Stable
Maturation time	Short	Long
Lifespan	Short	Long
Death rate	Usually high	Usually low
Number of offspring produced per reproductive episode	Many	Few
Number of reproductions in lifetime	Usually one	Often several
Timing of first reproduction	Early in life	Later in life
Size of offspring or eggs	Small	Large
Parental care	None	Often extensive

It is difficult, though, to place organisms discretely into one of the two groups. There are many exceptions, and most sit somewhere along the scale between r-selected and K-selected. For example, trees can live for many years, in some cases hundreds, but produce massive numbers of offspring, thus displaying traits of both. Similarly, reptiles, such as sea turtles, display both r- and K-traits; although sea turtles are large in size, with very long life spans (provided they reach adulthood), they produce large numbers of offspring that receive no parental care. Furthermore, mammalian males tend to be r-type reproducers due to copious numbers of sperm, whereas females tend to have K-characteristics, producing far fewer eggs. There are many examples where organisms do not fit neatly into either category.

Life histories vary greatly from species to species. Pacific salmon spawn in small streams, then migrate to open ocean where they feed and grow for a number of years. Eventually, they journey back upstream to their spawning grounds, release huge numbers of gametes in a single reproductive event and then they die. There are clear characteristics here of r-selected species. In contrast, some lizards reproduce every year for a number of years and only lay a few large eggs.

Costs and benefits of external and internal fertilisation

Internal fertilisation is the process by which the sperm and egg nuclei fuse inside the female. Conversely, external fertilisation is when this fusion occurs outside the female. In external fertilisation, large numbers of eggs and sperms are released into the water and fertilisation occurs in the absence of parents. External water is essential here to prevent drying of the gametes and to provide a medium for sperm to swim to eggs. Courtship behaviour is important in external fertilisation because timing is crucial to ensure that mature sperm and ripe eggs meet. Environmental stimuli, such as temperature or day length, may cause all the individuals of a population to release their gametes at the same time, while chemical signals from one individual releasing gametes may trigger gamete release in others.

Internal fertilisation is necessary where no external moisture is available. Cooperative behaviour which ultimately leads to copulation is required. Internal fertilisation requires highly advanced reproductive systems.

External fertilisation in frogs by http://www.flickr.com/people/63048706@N06, licenced under the Creative Commons https://creativecommons.org/licenses/by/2.0/deed.en license

This frog carries the tadpoles on her back until she finds water by http://flickr.com/people/19731486@N07, licenced under the Creative Commons https://creativecommons.org/licenses/by/2.0/deed.en license

Protection and care

All species produce more offspring than survive to reproduce. Species with external fertilisation usually produce enormous numbers of gametes, but the proportion that are successfully fertilised, survive and develop further is often quite small. Internal fertilisation usually produces fewer offspring; however, success is usually greater. Embryos are offered greater protection and parental care is high. Types of protection that require a greater input of energy include:

- resistant eggshells;
- development of the embryo within the female parent;
- parental care of eggs and offspring.

© HERIOT-WATT UNIVERSITY

TOPIC 4. SEX AND BEHAVIOUR

Birds and reptiles produce eggs with calcium and protein shells that can withstand cold, dry, hot environments with rough terrain. In comparison, the eggs of fish and amphibians have only a gelatinous coat.

In mammals, embryos develop within the uterus, obtaining nourishment from the mother's blood supply via the placenta.

When a bird hatches or a mammal is born, it is still not capable of independent existence. Adult birds endure the demands of feeding their chicks and mammals nurse their offspring with milk produced by the mother. There are, however, endless examples of parental care, including some rather peculiar methods, all with the ultimate goal of increasing the chance of survival of the offspring. In one species of tropical frog, for example, the father carries the tadpoles inside his stomach until they hop out as frogs!

Optimal reproduction is based on the premise of a trade-off in terms of the number and quality of current offspring versus potential future offspring. Some plants and animals, such as desert annual plants, many insects and Pacific salmon, invest most of their energy in maturation, expending this energy in one single reproductive event before dying. Other organisms produce fewer offspring at a time over a number of successive seasons. The relative advantage of each strategy can be thought of in terms of a trade-off between fertility and survival probability. If an organism is to breed over a number of successive seasons, it must invest some energy in survival mechanisms. Unfortunately, if the organism dies before reproducing again, these resources have been wasted. A general rule is that organisms with a low survival chance between reproductive events will have one reproductive event, whereas those that survive well once established will have more than one, smaller reproductive events. Organisms inhabiting unstable, unpredictable or harsh environments will likely have one large reproductive event.

Sexual investment: Activity Go online

Q1: Complete the table using the attributes listed.

Characteristic	r-selected population	K-selected population
Environment		
Lifespan		
Number of offspring per reproductive episode		
Number of reproductions in lifetime		
Size of offspring or eggs		
Parental care		

Attributes: few, large, long, many, none, often extensive, often several, short, small, stable, unstable, usually one.

4.2 Reproductive behaviours and mating systems in animals

Polygamy versus monogamy

In many species, mating is promiscuous, with short-lasting relationships. In species where bonds are formed, the relationship may be described as monogamous (one male mating with one female) or polygamous (an individual of one sex mating with several of the other). Polygamous relationships most commonly are between a single male and many females.

Most birds choose **monogamy** due to the high level of parental care needed in feeding the chicks. It is unlikely that one parent could meet these feeding and protecting requirements alone. In a monogamous relationship, male birds probably have a better chance of leaving more viable offspring by participating in parental care than searching for other mates. In mammals, the female is often the only food source for the young by means of the milk that she produces. Males may have a role in protection, but not feeding, therefore **polygamy** is common here with one male mating with many females in a harem, and protecting them and his offspring, as seen in deer and lions.

In some mating situations, one male may mate exclusively with a group of females. This is know as polygyny. Polyandry however, is the reverse - one female mates with a number of males in the same breeding season.

Another factor that influences reproductive strategy is certainty of paternity. The female who gives birth or lays the eggs is certain of maternity. However, even within a normally monogamous relationship, the male cannot be 100% sure of paternity.

A mammal with an insect's strategy

A particularly unusual reproductive strategy is employed by the naked mole-rat. Naked mole-rats are indigenous to the hot, dry areas of Kenya and Ethiopia. Like social insects, they live in underground colonies numbering between 100 and 300 individuals, and only a single 'queen' reproduces with one or two males. The rest of the colony assume the role of 'worker' or 'soldier'. Only 2% of the colony are fertile at any one time. Infertility of the others can be reversed if the queen dies. This reproductive strategy seems to have evolved in naked mole-rats due to the harsh conditions in which they live. Low rainfall, lack of food and hard ground means that living in social groups in shared burrows will increase survival in such an inhospitable environment. Due to genetic relatedness, the fitness of non-breeding individuals will be far greater within a colony than it would be with a solitary existence.

Naked mole-rat in its burrow

Sexual dimorphism

Sexual dimorphism is defined as any physical differences between males and females. It is often expressed as a difference in size with the male usually being larger, but it also involves such features as colourful plumage in male birds, manes on male lions, antlers on male deer and other adornments. In most cases of sexual dimorphism, the male is conspicuous and the female very inconspicuous. It makes sense for the female to be inconspicuous because she may carry and protect young, so an ability to camouflage or blend in will increase her and her offspring's chance of survival. In some cases, the male with the most impressive masculine features may be the most attractive to females. Secondary sex structures may also be larger, allowing males to compete with other males for females, e.g. antlers. These structures are often used for show, in an effort to appear threatening without the need for violence.

Darwin considered sexual dimorphism to be a product of sexual selection. This was because many of the features displayed in males, such as the plumage of peacocks, do not confer an advantage to survival in their environment, and in some cases can attract predators or merely get in the way. If these structures increase the individual's chance of gaining a mate, however, they will be favoured because they enhance reproductive success. In the case of male peacocks, a beautiful plumage will demonstrate fitness and imply a low parasite burden - attractive characteristics to a female.

Grand plumage of the peacock   *Male lion characterised by its mane*

In most species, females are very choosy; picking a poor-quality male can be a costly error. Males must therefore win at least one, but in most cases many females. In some animals, competition or rivalry among males almost entirely determines which animals will mate. In other species, females assess mates based on specific behaviours which she observes in the male, e.g. keeping a tidy, clean nest or foraging for sufficient quantities of food of a good quality. Such behaviours are important because they provide the female with an idea of how much parental care the male may offer. For example, male common terns (which are similar to gulls) carry fish and display them to potential mates as part of the mating ritual. Eventually, a male may begin to feed fish to a female. This behaviour conveys a male's ability to provide food for chicks. In other species, females choose males who exhibit extreme and energetic courtship displays or who have the most extreme secondary sex characteristics, such as a long tail. Perhaps these features indicate that the male is vigorous and in good health, thus a sign of fitness. Fitness can be defined in terms of good genes and low parasite burden.

Lekking

In a number of bird and insect species, males display communally in a small area called a **lek**. Females visit the lek and choose among displaying males by assessing whether signals are 'honest' and are really a sign of male fitness. They will look for good condition and low parasite density, both implied by a healthy plumage. A healthy plumage can be used to infer that the male has strong genes against parasites and diseases, clear traits that the female would want for her offspring

Black Grouse in a lek by http://www.flickr.com/people/41502344@N02, licenced under the Creative Commons https://creativecommons.org/licenses/by/2.0/deed.en license

After females makes their choice, mating occurs, after which there is no further contact between females and males.

Alternative mating strategies - sneaking behaviour in males

Sneaking behaviour refers to a strategy that allows smaller, less threatening males to more stealthily access a female partner, often avoiding altercations with dominant males. These sneaking males are often called **satellite males**.

Large horned beetles develop very large horns that they use as weapons for fighting for females. Smaller horned beetles do not possess such weaponry so adopt alternative mating strategies. Beetles with long horns will guard their mate by protecting the entrance to the tunnel. These males will fight any male that tries to enter. Smaller males with little or no horns have little chance of beating larger males so opt for a sneaking strategy. These satellite males dig a new tunnel that allows them access to the female's tunnel, unnoticed by the male guard. Since both of these strategies have proven, thus far, to be reproductively effective for the males that practice them, male horned beetles continue to have two very different, yet successful, phenotypes.

Rhinoceros beetle with large horn   *Bluegill sunfish*

Sneaking behaviour is also common in many fish species. Male Bluegill sunfish come in three different size morphs.

1. A large male that courts females, and then defends a nest in which he rears young.

2. A medium sized satellite or sneaker male that successfully fertilises eggs by mimicking the females in order to interrupt courtship between the female and a large male.

3. A very small satellite that dives in between a mating pair and squirts ejaculate in order to fertilise eggs.

All three mating types mentioned above are considered by researchers to have approximately equal fitness.

Species with larger females than males

In some species there may be reversed sexual dimorphism where the female is more conspicuous. This is common amongst insects, spiders, fish, reptiles and birds of prey. There are some cases in mammals, e.g. spotted hyenas and blue whales. Males are often smaller in species where they may have long distances to travel in search of females. Obviously, a smaller size will be an advantage for speed and stealth.

One example of sexual size dimorphism is the black myotis bat. Females are significantly larger than males in body weight, skull size and forearm length. Females are thought to be larger because they have the higher energetic costs of producing eggs in comparison to the 'cheaper' lower energetic costs faced by males in sperm production. Larger mothers are an advantage in organisms where gestation and lactation last for quite a number of months, such as the black myotis bat where females suckle their offspring until nearly adult size. They would not be able to continue flying and catching prey if they were not larger, thus able to support the additional mass of the offspring during this time.

Some species of angler fish display extreme reversed sexual dimorphism. Females are much larger than males. Males live almost a parasitic existence with an underdeveloped digestive system. Upon finding a mate, the male fuses with her, embarking upon a parasitic existence and becoming little more than a sperm-producing body.

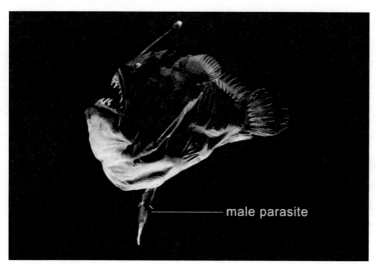

Female angler fish with the smaller male attached by Edith Widder / EOL, licenced under the Creative Commons https://creativecommons.org/licenses/by-nc-sa/3.0/ license / "male parasite" label added

There are many more examples of reversed sexual dimorphism in nature.

Species-specific signals

Animals often use signals that only other members of the same species understand. These may include giving off chemicals, or making sounds or certain displays to initiate mating. This means that individuals mate when most fertile, increasing the chance of successful reproduction. In birds and fish, in particular, successful courtship behaviour can be a result of species-specific sign stimuli and the resulting fixed action pattern responses. These signs and signals will be innate, thus instinctive.

A good example of fixed action patterns has been observed in male three-spine sticklebacks. During mating season, the male will develop a red colour on his throat. This sign stimulus attracts females and entices aggression in other males. The male builds a nest, attacking any males that try to enter, whilst courting and attracting females. Using stickleback models, researchers discovered that males respond aggressively to the red throat stimulus. Conversely, when a model with a swollen stomach was used, males responded by demonstrating courtship behaviour. These fixed action patterns will increase fitness by increasing the chance of successful mating.

TOPIC 4. SEX AND BEHAVIOUR

Male three-spined stickleback with red throat by Piet Spaans, licenced under the Creative Commons https://creativecommons.org/licenses/by-nc-sa/3.0/ license

Courtship: Questions Go online

Q2: The difference in phenotypic appearance between males and females of a species is a definition of:

a) lekking.
b) sexual dimorphism.
c) sneaking behaviour.

....

Q3: The process by which male birds display in a communal area so that females can assess and choose a mate is a definition of:

a) lekking.
b) sexual dimorphism.
c) sneaking behaviour.

....

Q4: A smaller male that gains access to females without other dominant males knowing is a definition of:

a) lekking.
b) sexual dimorphism.
c) sneaking behaviour.

© HERIOT-WATT UNIVERSITY

The following table shows how parasite infestation affects mating success in male capercaillies. Mating success is measured as how many times a male is selected by a female to mate during the breeding season.

Parasite abundance	Average number of times the male is selected by a female
Low	35 +/- 1.5
Medium	15 +/- 1.5
High	2 +/- 1.5

Q5: Present the information in the table as a bar graph.

....

Q6: Calculate the percentage change in the number of times a male is selected when parasitism rises from low to medium.

....

Q7: What conclusion can be drawn about parasite abundance and mating success?

....

Q8: Suggest why parasite abundance is a useful way for females to assess males.

4.3 Learning points

Summary

- Egg production, with their energy store and larger size, involves a far greater energy investment than sperm production, meaning that females have a greater sexual investment.
- Organisms will employ the most successful reproductive strategy possible, assessing the number and quality of current offspring against the potential for future offspring.
- Organisms can be classed as r-selected or K-selected organisms based on their reproductive strategy and life history, although organisms can display traits of both groups.
- Sexual dimorphism is when males and females of the same species have very different physical appearance, often in terms of colour and size - this is a result of sexual selection.
- Females are usually inconspicuous.

> **Summary continued**
> - Sexual dimorphism can be reversed in some species.
> - Male-male rivalry sees larger males, often with appendages used a weapons, having greater success at finding a mate.
> - Smaller males may employ sneaking behaviour and see equal success at gaining access to females.
> - Successful courtship is often the result of species-specific sign stimuli that bring about a fixed action pattern response.
> - Lekking is when males of a species, usually birds, gather in a display area called a lek; females attend the lek and choose a mate.
> - Females assess male fitness, basing it on 'honest' signals such a low parasite burden and good condition.

4.4 Extended response question

The activity which follows presents an extended response question similar to the style that you will encounter in the examination.

You should have a good understanding of courtship before attempting the question.

You should give your completed answer to your teacher or tutor for marking, or try to mark it yourself using the suggested marking scheme.

> **Extended response question: Courtship**
>
> Write an essay about courtship under the following headings:
>
> - sexual diamorphism; *(4 marks)*
> - male-male rivalry; *(1 mark)*
> - sneakers; *(1 mark)*
> - lekking. *(4 marks)*

4.5 End of topic test

End of Topic 4 test

Q9: Sexual investment in females is _____ that in males. *(1 mark)*

a) equal to
b) greater than
c) lower than

....

Q10: Which of the following are characteristics of r-selected organisms? *(1 mark)*

a) Production of few offspring, short life spans, parental care.
b) Production of few offspring, long life spans, no parental care.
c) Production of many offspring, short life spans, no parental care.
d) Production of many offspring, long life spans, parental care.

....

Q11: What is the term that is used to describe the condition where there are distinct physical differences between males and females of the same species? *(1 mark)*

....

Q12: Which of the following statements best explains why females are often less conspicuous than males? *(1 mark)*

a) To reduce the chances of their being confused as males by other females.
b) To provide better camouflage when protecting their young.
c) To reduce the chances of males trying to fight them.
d) Due to a lack of energy required to look nice after producing 'costly' eggs.

....

Q13: How can males that are smaller or lack weapons still gain access to females? *(1 mark)*

Unit 2 Topic 5

Parasitism

Contents

5.1 Niche 198
5.2 Parasitic life cycles 202
5.3 Transmission and virulence 210
5.4 Defence against parasitic attack 211
    5.4.1 Non-specific defence 211
    5.4.2 Specific defence 213
5.5 Immune evasion 216
5.6 Challenges in treatment and control 217
5.7 Learning points 220
5.8 Extended response questions 222
5.9 End of topic test 223

Prerequisites

You should already know that:

- parasites form symbiotic relationships with hosts, where the parasite benefits at the host's detriment;
- ectoparasites, such as ticks, live on the surface of a host as compared to endoparasites, such as tapeworms, which live inside the host;
- more highly evolved parasites have indirect life cycles;
- parasites can be transmitted by vectors;
- interspecific competition occurs between organisms of two different species and can be reduced through compromising over resources;
- some methods by which parasites are transmitted include direct contact, consumption of secondary hosts and vectors;
- malaria is an example of a human disease caused by a parasite;
- DNA codes for proteins;

Prerequisites continued

- during the process of protein synthesis, DNA is transcribed into a molecule of mRNA which is then translated into a polypeptide strand that is further modified into a functional protein;
- the body may defend attack using non-specific and specific means;
- white blood cells provide cellular defence against foreign antigens;
- antibodies are special proteins produced during immune responsees;
- vaccines and drugs are being used and continue to be developed to treat illness;
- overcrowding and poor sanitation are linked to rapid spread of parasites.

Learning objective

By the end of this topic, you should be able to:

- describe the symbiotic relationship shared by parasites and their hosts;
- state what is meant by a narrow niche as applied to parasites;
- explain what degenerate means as applied to parasites;
- define the terms definitive host and intermediate host;
- explain the term vector using examples;
- compare the fundamental and realised niche of parasites;
- explain what is meant by competitive exclusion principle and resource partitioning;
- describe the relationship between parasite transmission and virulence;
- give examples of factors that increase transmission rates of parasites;
- explain what is meant by the theory 'The Extended Phenotype';
- give examples of how parasites modify their hosts for their own gain;
- state, with examples, that many parasites require more than one host to complete their life cycle;
- give examples of parasites that are responsible for causing human diseases;
- give examples of parasites that can complete their life cycle within one host;
- give examples of human diseases caused by bacteria and viruses;
- state what a virus is and describe its structure;
- state that the outer surface of a virus contains antigens that a host cell may or may not be able to detect as foreign;
- explain the role of the lipid membrane envelope around some viruses and state what it is derived from;

> **Learning objective continued**
> - describe the process of virus replication;
> - define the term retrovirus;
> - explain how retroviruses replicate compared to viruses.
> - give examples of physical and chemical barriers in non-specific defence;
> - describe inflammatory response and explain its role in non-specific defence;
> - describe the process of phagocytosis and the role of phagocytes in this process;
> - state the role of natural killer cells in non-specific defence;
> - describe the role of phagocytes in specific defence;
> - compare the role of T and B lymphocytes in specific defence;
> - explain what clonal selection is and its role in immunological memory;
> - define the term epidemiology;
> - discuss how endoparasites can evade detection by the host's immune system;
> - state that vaccine design and drug development are both difficulties that need to be overcome in the fight against parasites;
> - state that culturing some parasites in the laboratory is difficult and may act as a barrier to learning how to control them;
> - explain that rapid change in parasite antigens adds to the challenge of vaccine and drug development;
> - state that host and parasite metabolisms being so similar adds challenges to finding treatments that only target the parasite;
> - state that improving sanitation and controlling vectors will help control parasites;
> - explain why parasites are more prevalent and difficult to control in developing countries, tropical climates or areas affected by natural disasters;
> - understand that controlling parasites will reduce child mortality in developing countries and that this will result in population-wide improvements.

5.1 Niche

An ecological niche is a multi-dimensional summary of tolerances and the requirements of a species. There are two ideas of 'niche' that we use with species, a fundamental niche and a realised niche.

At least 50% of all species are parasitic, and all free-living species are thought to host parasites to some extent. A parasite forms a symbiotic relationship with the host, benefiting at the expense of the host. The parasite gains nutrients and shelter from the host. Unlike other symbiotic relationships, such as those between predator and prey, the reproductive potential of the parasite is much higher than that of its host.

An ecological niche is a complex outline of tolerances and requirements of a species. Parasites tend to have a very narrow niche due to high host specificity. As the host provides so many of the parasite's needs, many parasites are said to be degenerate, lacking in structures and organs found in other organisms. A good example of this is tapeworms, which lack a digestive system as the host has already digested the food. This degeneration is also found amongst the flatworm parasites. Flatworms such as *Fasciola Ascaris* lack organs for movement, digestion of food and sensing light as these are not required whilst inside the host. This does limit parasites, in the sense that the adult parasite will not be able to survive outside the host.

The niche for an ectoparasite is on the surface of its host, e.g. lice, ticks and leeches, whereas an endoparasite lives within the host, e.g. tapeworms, flatworms and protozoans. There are two types of host.

1. The **definitive (primary) host** where the parasite reaches sexual maturity.
2. The **intermediate (secondary) host** which the parasite might require in order to complete its life cycle, perhaps carrying out asexual reproduction to greatly increase it numbers or as a means of **transmission**, thus sometimes making the intermediate host a **vector**.

A good example of a parasite that has two hosts is the tapeworm *T. solium*, a parasite found in human intestines. Humans act as the definitive host in this case. *T. solium* has evolved to specialise and adapt to two different hosts, which is difficult and highly advanced. The advantage is that large parasites, such as the tapeworm *T. solium*, which would not be able to leave their definitive human host easily can do so via their much smaller eggs. The eggs enter pigs if they eat food contaminated with human faeces, and can then be transmitted to a new human host via undercooked pig meat. The tapeworm may further increase its numbers by carrying out asexual reproduction within the intermediate (secondary) pig host. In CfE Higher Biology, we referred to this more evolved parasite life cycle as indirect.

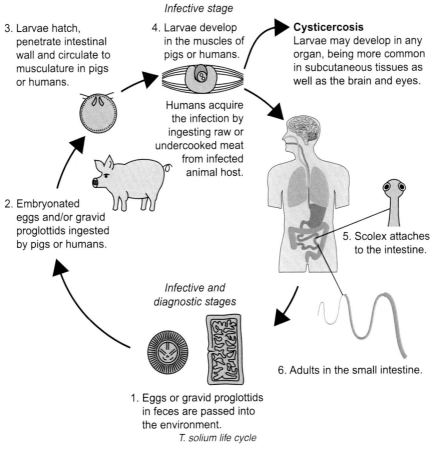

T. solium life cycle

In the image above, **gravid proglottid** refer to any segment of a tapeworm containing both male and female reproductive organs and **cysticercosis** refers to a parasitic tissue infection.

Vectors

Vectors are responsible for transmitting a parasite from one host to another, yet may also be a host for the parasite. Good examples of vectors include mosquitoes in transmitting the malaria parasite or ticks in transmitting both Lyme disease and babesiosis disease-causing parasites. Many endoparasites acquire hosts passively. The human endoparasitic nematode *Ascaris* produces vast quantities of eggs in the digestive tract that pass out in faeces. More people become infected through ingesting the eggs if there is poor sanitation and contamination of water supplies. Natural selection favours those parasites that can find and obtain nutrition from its host quickly and efficiently.

Washing clothes in or drinking this water in Tanzania will promote transmission of parasites by http s://www.flickr.com/people/23116228@N07, licenced under the Creative Commons https://creativec ommons.org/licenses/by/2.0/deed.en license

Ectoparasites and many endoparasites have sophisticated methods for finding new hosts. Mosquitoes, for example, initially try to locate a human host for the malaria parasite by sensing movement and then verifying that the host is human based on body temperature, exhaled carbon dioxide and chemicals on the surface of the skin.

In ecology, two niches exist.

1. The **fundamental niche** - this is the niche that the organism occupies when there are no other species present competing for space or resources.
2. The **realised niche** is the niche that the organism occupies when there is competition from other species.

Competitive exclusion principle is when two species are in intense competition with one another, thus the niches occupied by each are very similar - the weaker of the two species will likely die out, becoming extinct in that area.

Resource partitioning is where two competing species occupy different realised niches, compromising over resources and thus managing to exist simultaneously. These terms and concepts can be applied to parasite niches as well as any other species.

Nematode species	Section of the intestine inhabited (%)
Passalurus ambiguus	8-25
Obeliscoides cuniculi	19-37
Trichuris leporis	30-52
Dirofilaria scapiceps	45-68
Bayliascaris procyonis	62-88

Range of rabbit intestine occupied by five different nematode species

The table shows five different species of nematode parasite found in the intestines of rabbits. To reduce competition, each has specialised to position itself at a certain area of the intestinal tract. For example, *Passalurus ambiguus* inhabits the first 8-25% of the intestine. It overlaps slightly with *Obeliscoides cuniculi*, which can be found from 19% along the tract to 37%. This will involve adapting to the different environmental conditions, such as nutrient availability and pH level. They are said to have effectively carried out resource partitioning.

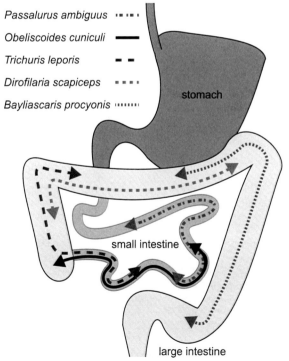

Fundamental and realised niches of nematodes in intestines

The parasite niche of Clostridium difficile

C. difficile is a bacterial species that lives in the large intestine of humans and is transmitted via spores produced and spread in faeces. Initial symptoms include very bad diarrhoea. High risk patients are those on antibiotics who are currently under medical care in hospitals, nursing homes, outpatient surgery centres or health centres. Antibiotics are used to kill bacterial infections. It is therefore unavoidable that they kill the harmless bacteria with which we share a positive symbiotic relationship, including those that live in the gut. This leaves a niche open for harmful *C. difficile* bacteria to fill.

Niche: Activity Go online

Q1: Match the terms with their definitions.

Term	Definition
Fundamental niche:	two different species compromise over resources to reduce competition.
Realised niche:	competition between two species will see local extinction of the weaker.
Resource partitioning:	required by some parasites to complete their life cycle.
Competitive exclusion principle:	where the parasite reaches sexual maturity.
Vector:	exists in the absence of interspecific competition.
Definitive host:	exists in the presence of interspecific competition.
Intermediate host:	responsible for parasite transmission.

5.2 Parasitic life cycles

A parasite is a symbiont that gains benefit from its host by using its nutrients. Some parasites will only require one host to complete their lifecycles but many will require more than one host. This is why vectors play an important role in the transmission of parasites that require more than one host - they aid transmission. The vector itself, for some parasites, may also become a host in certain stages of a life cycle.

The human disease malaria is caused by *Plasmodium* spp. The parasite will enter the human bloodstream to asexually reproduce in the liver and then in red blood cells. afer the red blood cells bursts, gametocytes will be released into the humans bloodstream and another mosquito biting this particular infected human gains the gametocytes. Sexual reproduction can now occur when the gametocytes mature into female and male gametes. The mosquito is now ready to infect another host.

Malaria is a serious tropical disease spread by mosquitoes that are infected with the malaria parasite *Plasmodium* of which there are several different species. Presenting as a headache and fever, this is treatable if detected early enough. If treatment is not strong enough, malaria can recur. Sometimes, the parasite can exist in a dormant state, often hiding in the liver cells so is undetected by the body's immune system, allowing future recurrence. In the *Plasmodium falciparum*, the parasite has adhesive proteins on its surface, allowing it to stick to the inner wall of the blood vessels. This also allows it to remain undetected by the immune system.

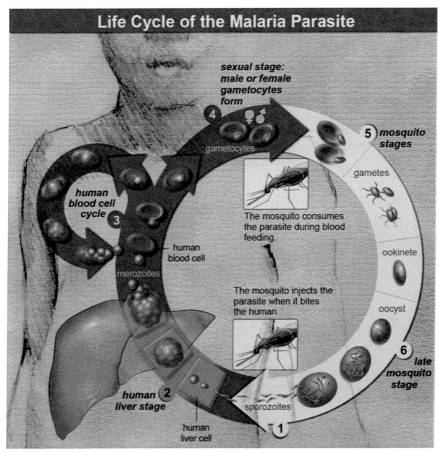

Malaria life cycle

Schistosomiasis are also known as bilharzias. This disease is caused by flatworms of the *Schistosoma* species which live in fresh water in sub-tropical and tropical regions of the world. Presenting as a skin rash and cough, this is readily treatable in the UK.

Schistosomes reproduce sexually in the human intestine. The fertilised eggs pass out in faeces into water where they are able to develop in to larvae. water snails are then infected by the larvae and asexual reproduction can occur. the result is another type of motile larvae which can escape from the snail and penetrate human skin to enter the bloodstream.

Schistosomiasis

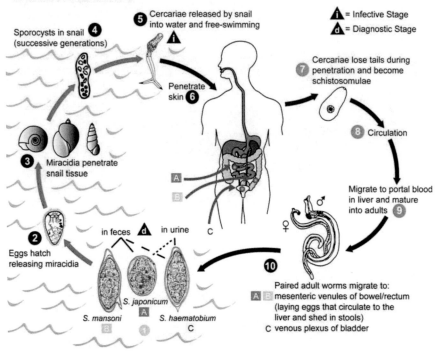

Schistosomiasis life cycle

Both of these parasites require more than one host to complete their life cycle. In the case of malaria, the parasite grows and multiplies inside the human host. The female mosquito acts as a second host, not only responsible for transmitting the parasite, but also for allowing growth and multiplication of the parasite when in a stage known as the blood stage. *Schistosoma* flatworms also have two hosts: some freshwater snails and humans. Infectious parasites leave the snail host, entering freshwater. Any human swimming in infected water may become infected when the skin comes in contact with the water.

Other human diseases can complete their life cycle within one host.

1. Ectoparasitic arthropods, e.g. ticks.
2. Endoparasitic protists, e.g. amoebas.
3. Bacteria, e.g. tuberculosis.
4. Viruses, e.g. influenza and HIV.

Parasitic life cycles: Questions

Go online

Q2: A malaria parasite which requires both human and mosquito hosts to complete its life cycle is:

a) an arthropod tick.
b) *Plasmodium*.
c) *Schistosoma*.

....

Q3: A platyhelminthes parasite which requires both human and freshwater snail hosts to complete its life cycle is:

a) an arthropod tick.
b) *Plasmodium*.
c) *Schistosoma*.

....

Q4: A parasite with jointed legs, segmented body and exoskeleton, capable of completing its life cycle on a single host is:

a) an arthropod tick.
b) *Plasmodium*.
c) *Schistosoma*.

The unicellular parasite *Trypanosoma brucei* attacks the human nervous system causing sleeping sickness which is a common condition in regions of the world such as Africa where tsetse flies are found. *Trypanosoma brucei* contains variant surface glycoprotein coats which are so thick that the immune system of the host cannot access the plasma membrane of the parasite. In addition, the parasite undergoes regular antigenic variation. The following flow chart shows the life cycle of *Trypanosoma brucei*.

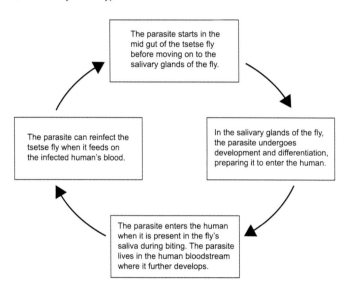

Q5: What parasite category does *Trypanosoma brucei* belong to?

....

Q6: Name the two hosts of *Trypanosoma brucei*.

....

Q7: Using the information provided, explain why humans are unable to fight off the parasite without medication.

....

Q8: Current medicines can prove fatal to the human host, therefore research into new medicines and vaccines is essential. Using the information provided, explain why this is proving difficult.

Viruses

Viruses are tiny (20-400 nm) infectious agents that can only replicate inside a host cell. Viruses contain genetic material in the form of DNA or RNA, packaged in a protective protein coat as shown in the picture of the **bacteriophage**. A bacteriophage is a virus that uses a bacterium as a host cell. The outer surface of a virus contains antigens that, in some cases, the host cell may not be able to detect as foreign.

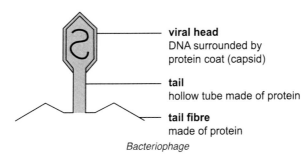

Bacteriophage

Lipid membrane

The protein coat of some viruses is surrounded by a lipid membrane envelope. This envelope is composed of the host cell's materials. It is thought that the envelope has a role in allowing the virus to attach to the host and gain entry. It may also help the virus avoid detection by the host cell.

Replication of a virus

Viral replication has the following steps.

1. Virus attaches to the surface of the host cell.
2. Virus injects its DNA or RNA into the host cell.
3. Virus interrupts the host cell's own metabolism, often entering the host's genome.
4. Virus uses the host cell's machinery and raw materials to replicate the DNA.
5. Again, using the host cell's resources, the DNA is transcribed into mRNA then translated into protein so more viral protein coats are produced.
6. The new DNA then enters the newly formed protein coats, thus producing many new viruses.
7. These then leave the cell to infect new cells and the host cell undergoes **lysis**, bursting.

Viral replication

- viral particle
- virus adheres to the host bacterium cell
- virus injects its DNA into the host bacterium
- bacterium

- host cell's metabolism is interrupted
- viral DNA replicates using bacterial enzymes and nucleotides

- viral DNA is transcribed into mRNA and translated into viral protein coats by the bacterium

- protein coats surround viral DNAs

- host bacterium breaks up (lysis)

- viral particles released to infect othe cells

Retroviruses

Retroviruses are viruses with RNA rather than DNA. HIV is a good example of retrovirus. On injecting their RNA into the host cell, they also inject the enzyme **reverse transcriptase** to first synthesise DNA from the single-stranded RNA. This new DNA is then inserted into the genome of the host cell. As part of the host's DNA, the virus's genes can then be transcribed, ultimately synthesising new viral particles.

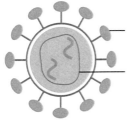

glycoprotein on membrane envelope
binds receptor on T helper cell

capsid
contains two single-stranded RNA molecules and two molecules of the enzyme reverse transcriptase

HIV virus structure

A glycoprotein on the surface of the HIV envelope binds to a specific receptor on the surface of a helper T cell. Leaving its envelope behind at the cell membrane, the viral particle enters the host T cell. The capsid proteins are removed by enzymes. Viral reverse transcriptase catalyses the synthesis of a DNA strand complementary to the viral RNA template. A complementary DNA strand is synthesised and the double stranded DNA is incorporated into the host's genome. Viral genes are transcribed into mRNA to make new viral genomes. The mRNA is also translated into HIV proteins in the cytoplasm. New viral particles are assembled. As each viral particle buds out of the host cell it is coated by a membranous envelope.

Retroviruses and evolution in eukaryotes

The genomes of most eukaryotic species contain a high proportion of **retrotransposons**. Retroviruses are thought to be the origin of these retrotransposons. The great variability of vertebrate antibodies is hypothesised to have evolved from retrotranposons.

Parasitic life cycles: Activity   Go online

Q9: Complete the diagram using the labels listed.

RNA	→		→	
-		-		-

Label list: **DNA**, host RNA polymerase, **RNA**, viral reverse transcriptase.

5.3 Transmission and virulence

Transmission is the spread of a parasite to a host. **Virulence** is the deleterious effect that the parasite has on the host.

Ectoparasites are normally transmitted through direct contact or by the consumption of intermediate hosts whereas endoparasites of the body tissues are more often transmitted by vectors.

The rule is that, the higher the transmission rate, the greater the virulence.

Transmission rates are greatest when there is:

- overcrowding of hosts, such as in places with very high population density, e.g. cholera;
- a means of transmission such as presence of **vectors**, e.g. mosquitoes or water for **waterborne** parasites or waterborne dispersal stages.

Host behaviour can often be exploited and modified by parasites so that they can maximise their transmission. Parasites will suppress the host immune system and modify their host size and/or reproductive rate to benefit parasitic growth. This is referred to as the host behaviour becoming apart of the extended phenotype of the parasite.

Transmission and virulence: Questions   Go online

Virulence refers to the percentage of infections that result in death. The following table presents information about the number of cases of the disease toxocariasis in foxes and the number that resulted in the death of the fox. This disease has been monitored over recent years because it is the most likely disease affecting foxes that could be transmitted to humans.

Year	Number of reported toxocariasis cases	Number of infections resulting in fox death
2010	13	8
2011	56	50
2012	144	75
2013	95	70
2014	25	15

Q10: Calculate the percentage change in number of reported cases from 2011 to 2012.

....

Q11: In which year was toxocariasis most virulent?

a) 2010
b) 2011
c) 2012
d) 2013
e) 2014

Q12: In which year was toxocariasis least virulent?

a) 2010
b) 2011
c) 2012
d) 2013
e) 2014

Q13: Summer 2011 was particularly warm with plenty of rabbits and small rodents available for foxes to feed on. Knowing what you do about the factors that increase parasite transmission, suggest why 2012 saw the greatest number of reported toxocariasis cases.

5.4 Defence against parasitic attack

Defence against parasitic attack comprises non-specific defence and specific defence.

5.4.1 Non-specific defence

Immune response to parasites

Mammals have innate or natural, non-specific defences to parasites (which includes disease causing bacteria and viruses). Physical barriers include:

- skin;
- chemical secretions, including mucus, tears and stomach acid;
- **inflammatory response**;
- **phagocytes** (white blood cells);
- **natural killer cells** - lymphocytes (white blood cells) responsible for destroying abnormal cells.

Should this defence ultimately fail, mammals are armed with specific or adaptive cellular defence involving immune surveillance by white blood cells.

Inflammatory response

When the skin is damaged, perhaps by a cut or wound, the external barrier to parasites is broken. Parasites may enter and this triggers localised inflammatory response. The wounded region becomes warmer and redder as a result of small blood vessels dilating. This increases blood flow to the injured area, thus increasing the number of white blood cells, such as phagocytes, allowing a suitable defence or attack so that healing can begin.

Phagocytes

Phagocytes are white blood cells with a non-specific role in defence. Phagocytes arrive at the site of infection and engulf parasites by enfolding their plasma membrane around the parasite. The parasite is then brought into the phagocyte in a vacuole or vesicle. Phagocytes contain special organelles called lysosomes. These are filled with digestive enzymes. The lysosomes fuse with the vacuole, releasing the digestive enzymes and allowing the enzymes to digest the parasite. The following depicts the process of **phagocytosis**.

Phagocytosis Go online

Phagocytes also have a role in specific defence. Foreign antigens, previously engulfed by phagocytes, are pushed back out onto the surface of the phagocyte. The phagocyte then presents these antigens to lymphocytes, another type of white blood cell.

Phagocyte is attracted to chemical signals produced by a bacterium. 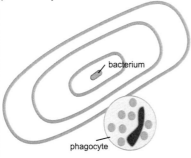	Vacuole forms around the bacterium. Lysosomes move towards and fuse with the vacuole. 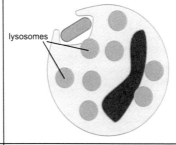
Lysosomes release digestive enzymes into the vacuole, the bacterium is broken down by enzymes. 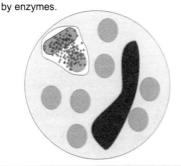	Vacuole disintegrates releasing digested products into the cytoplasm of the phagocyte. 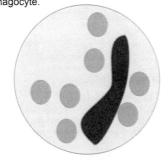

> **Defence against parasitic attack: Activity** Go online
>
> **Q14:** Arrange the following steps in phagocytosis in the correct order.
>
> - The parasite is digested.
> - Parasite is brought into the phagocyte in a vacuole.
> - Phagocytes move to the site of injury.
> - Lysosomes fuse with the vacuole releasing digestive enzymes.
> - Plasma membrane of the phagocyte engulfs the parasite.
> - Lysosomes move towards the vacuole.

5.4.2 Specific defence

Specific defence is carried out by a range of white blood cells. White blood cells carry out **immune surveillance** by constantly circulating to monitor the tissues. If tissues become damaged or invaded, cells can release cytokines that increase blood flow. This results in non-specific and specific white blood cells accumulating at the site of infection or where damage has occurred.

Lymphocytes

Lymphocytes are part of a specific response, with a different lymphocyte produced in response to each type of foreign antigen.

Mammals are armed with two types of lymphocytes:

- lymphocytes also known as B cells;
- lymphocytes also known as T cells.

Both types circulate throughout the blood, and recognise specific foreign antigens. Antigens are proteins on the surface of all cells. If these antigens do not belong to the host, they are considered foreign. These include foreign molecules belonging to bacteria, viruses, fungi, parasitic worms and cells from transplanted tissues.

In response to the foreign antigens presented by phagocytes, B cells produce proteins called antibodies that are specific in shape to the antigen.

T cells work by destroying specific infected or damaged cells by bringing about **apoptosis** (cell death).

Clonal selection

Lymphocytes carry out **clonal selection**. Clonal selection is process by which lymphocytes become amplified undergoing the process of mitosis. Each type of lymphocyte divides and differentiates into two clones. One clone becomes a plasma cell, working to combat the antigen and is therefore only short lived. In comparison, the other clone lives significantly longer assuming the role of a **memory cell**. Should the individual be exposed to a particular parasite a second time, lymphocyte memory cells very quickly mount a defence producing the specific antibody required to combat the antigen. In such cases the individual would express no symptoms whatsoever.

Clonal selection of lymphocytes

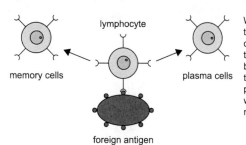

When a foreign antigen enters the body, the lymphocyte that contains the antibody receptor that is specific to the antigen binds to it. The lymphocyte then undergoes mitosis to produce many daughter cells, which then develop into either memory cells or plasma cells.

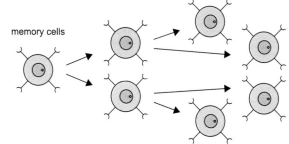

Having undergone mitosis, there are many copies of the antibody specific to the antigen in the body.

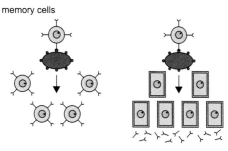

Memory cells recognise the antigen if it enters the body again and they initiate the secondary immune response. Since there are many memory cells in the body, antibody-secreting plasma cells can be produced rapidly to destroy the foreign antigen.

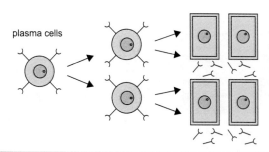

The lymphocytes that develop into plasma cells secrete antibodies in response to foreign antigens. These antibodies are specific to the antigen detected and destroy it.

Epidemiology

Epidemiology is the study of the outbreak and spread of infectious diseases. Diseases spread more quickly through dense populations and areas of overcrowding. In every population, as a result of genetic variation, some individuals will be genetically immune to a disease. Since diseases are transmitted from individual to individual, resistant individuals will present a barrier to transmission. If this number of resistant individuals is high enough, an epidemic may be avoided. These resistant individuals are ultimately providing a level of protection to non-immune members of the population.

Defence against parasitic attack: Questions   Go online

The following table provides data about immune response to chicken pox.

Time (days following infection)	Volume of chicken pox antigen (μg/ml)	Volume of antibody (μg/ml)
0	50	0
2	140.5	0
4	220	0
6	150	240
8	130.5	360
10	50	380
12	20	380

Q15: Present the information in the table as a line graph.
....

Q16: Use values from the table to describe the volume of chicken pox antigen over the 12 days.
....

Q17: Describe the volume of antibody over the 12 days.

The individual is exposed to chicken pox a second time. The following graph shows the levels of chicken pox antigen and antibody levels this time.

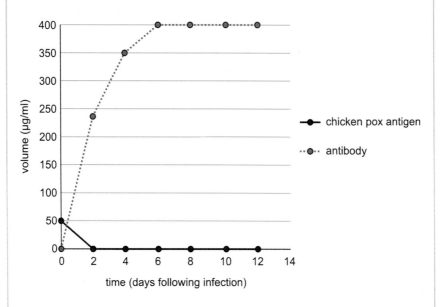

Q18: Explain the volumes of chicken pox antigens and antibody over the 12 days following infection.

5.5 Immune evasion

Endoparasite Success

So why don't endoparasites stimulate an immune response from mammals since their cells will contain foreign antigens?

Endoparasites mimic host antigens, therefore remaining undetected by the host's immune system. Endoparasites are also capable of modifying the host-immune response, preventing the host from mounting an attack so the parasite remains alive.

Some parasites show a huge **antigenic variation**. This promotes a rapid evolution rate, and the parasite remains a step ahead of host immune cell clonal selection. This is why flu is so successful. Flu has evolved over generations so that there are many forms, and being immune to one form does not guarantee immunity to other types.

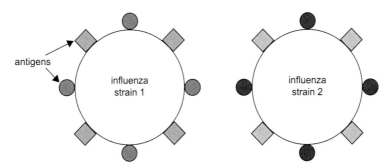

Compare the antigens of the two influenza strains here. Memory cells from strain 1 would provide no immune response to an attack from strain 2.

Antigenic variation in some parasites allows them to change between different antigens when they are infecting a host. This also means that the host can be re-infected by the same parasite using a new variant.

Some viruses will escape the immune surveillance by integrating their genome into the hosts. This allows them to exist in an inactive state that we call latency and will only become active again when favourable conditions arise.

5.6 Challenges in treatment and control

Epidemiology is the study of the outbreak of infectious diseases and their spread. The herd immunity threshold is the density of resistant hosts in a population required to prevent an epidemic.

There are many challenges to overcome in the successful treatment and control of parasites.

1. Some parasites are difficult to culture in the laboratory, so working with them to learn more about fighting them is a battle. The similarities between the host and parasite metabolism makes it difficult to culture parasites in a laboratory, offers difficulties in finding drug compounds that only target the parasite and makes it harder to design vaccines that reflect antigenic variations.

2. Many parasites show rapid evolution rates, often due to short generation times. This means that their **antigens** change quickly in structure. A good example is the influenza virus. There are many strains, each with different antigens. This means that designing vaccines for some parasites is almost impossible and can be expensive. In the case of influenza, UK scientists have been very successful in terms of designing vaccines and immunising many vulnerable individuals in the population. However, individuals must be vaccinated annually due to the rapid change in influenza antigen structure. For other parasites, there has been less success. Where research and development is too expensive, research into vaccines against particular parasites must be suspended.

3. Host and parasite metabolisms are extremely similar and intricately connected. Many parasites modify the host's metabolism and, in the case of viruses, their DNA becomes part of the host cell's genome. This means that tailoring drugs to fight the parasite without harming the host is extremely difficult.

Sanitation

In the UK, we have won the battle against certain parasites which other countries have not. We have **civil engineering** projects to thank for this, including sewage systems that use microbes to break down sewage into harmless products. This allows us to eradicate many diseases, such as cholera and dysentery, that are still common in poorer or less technologically advanced countries which have poor **sanitation**.

Vector Control

Another solution is coordinated vector control, killing the carrier, however this has proved controversial. In the 1950's, the pesticide DDT, also called Agent Orange, was used to kill malaria carrying mosquitoes. The disadvantage is that DDT is persistent and has subsequently bio-accumulated throughout food chains. The ultimate effect has been noted in bird of prey populations where, due to thinning eggs, numbers have fallen. Thinning eggs has been linked to the rather magnified levels of DDT in birds of prey that are in food chains containing mosquitoes.

Conditions that promote spread of parasites

Unfortunately, parasites spread most rapidly in those conditions where coordinated treatment and control programmes are most difficult to achieve. Often the most vulnerable people are affected, such as those who live in war torn regions or places affected by natural disasters.

- War - creates millions of refugees. People flee their homes and seek refuge in refugee camps. Camps become overcrowded and sanitation poor. This creates perfect conditions for rapid spread of parasites. Furthermore, reaching these refugee camps with resources to control parasites can be difficult if not impossible.
- Natural disasters - often destroy homes and sewage plants. Clean running water and sewage treatment cease, thus creating perfect conditions for spread of parasites.
- Tropical climates - parasites are more abundant in the tropical climates that are found in many developing countries. This is because tropical climates promote large populations of insect vectors that would need to hibernate during cooler seasons in colder climates.

Improvements in parasite control

As parasite control increases, child mortality decreases. Children are able to thrive in areas where parasites are under control. Thriving children have greater opportunities for growth, development and intelligence. More intelligent, healthy children will gradually become better educated, thus sustaining control of parasites into the future. Countries will only be able to sustain the fight against parasites if they have the education and tools, rather than relying on charities and support from developed countries.

Challenges in treatment and control: Questions

The following table provides data regarding four different countries.

Country	Sanitation	Climate	Population density (per km²)
1	Improved over recent years - basic sewage systems and running water.	Sub-tropical, with a high chance of hurricanes.	6200
2	Excellent - civil engineering strong.	Temperate, with some stormy winter weather.	2500
3	Very poor - water obtained from local rivers where clothes are washed.	Tropical, with high chance of typhoons.	7100
4	Education of local people has coincided with an improvement in sanitation - sewage facilities are being developed.	Sub-tropical, with seasonal heavy rains.	3000

Three hypotheses based on the above data are:

1. Country 1 will have the highest parasite abundance.
2. Country 2 will have the lowest parasite abundance.
3. Countries 3 and 4 have equal parasite abundance.

Q19: Do you agree with hypothesis 1? Use data from the table to support your opinion.

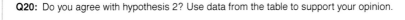

Q20: Do you agree with hypothesis 2? Use data from the table to support your opinion.

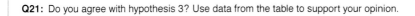

Q21: Do you agree with hypothesis 3? Use data from the table to support your opinion.

5.7 Learning points

Summary

- Parasites have a narrow niche due to high host specificity.
- Parasites are said to be degenerate due to the absence of certain structures and organs.
- Ectoparasites live on the surface of their host, whereas endoparasites live inside their host.
- The definitive or primary host is the host where the parasite reaches sexual maturity.
- In the case of indirect life cycles, some parasites also have a secondary or intermediate host that is used in transmission or where an asexual phase of their life cycle may occur.
- Indirect life cycles are highly evolved and greatly increase parasite success.
- Vectors, such as insects, or water transmit parasites.
- The fundamental niche is the parasite's niche in the absence of interspecific competition.
- The realised niche is the parasite's niche in the present of interspecific competition.
- When interspecific competition occurs between species with very similar niches, usually one of the two species becomes locally extinct.
- If the two species can compromise over resources by means of resource partitioning, both may co-exist.
- Higher rates of parasite transmission are linked to higher virulence.
- Transmission can be increased in overcrowded regions and where vectors are available. Parasites are said to modify host behaviour to increase transmission by altering host foraging behaviour, anti-predator behaviour, sexual behaviour, movement and habitat choice.
- These modifications are said to be an extension of the parasite's phenotype in a theory known as 'The Extended Phenotype'.
- Parasites also suppress host immune responses, and modify host size and reproduction for their own benefit.
- Schistosomiasis and malaria are examples of human diseases caused by parasites.
- *Plasmodium* species and *Schistosoma* species require two hosts to complete their life cycles.
- Other parasites can complete their life cycle within one host. Examples include endoparasitic amoebas and ectoparasitic ticks.
- Many bacteria and viruses can complete their life cycles within a single host.
- Viruses are composed of nucleic acid enclosed inside a protein coat.

Summary continued

- The outer surface of a virus contains antigens that a host cell may or may not be able to detect as foreign.
- Some viruses are further surrounded by a lipid membrane, composed of the host's materials.
- The lipid surround is thought to aid the virus in remaining undetected inside the host cell.
- Retroviruses are viruses that contain RNA as their genetic material rather than DNA. HIV is an example of a retrovirus.
- RNA retroviruses use the enzyme reverse transcriptase to first produce DNA, which is then inserted into the genome of the host cell. These virus genes form new viral particles when transcribed inside the host.
- Mammals are armed with non-specific defences to parasite attack. The skin provides a physical barrier, and mucus, saliva, tears and stomach acid all provide chemical barriers.
- Inflammatory response is a non-specific defences that increases the blood flow to an injured area.
- Phagocytes provide non-specific defence by engulfing foreign antigens into a vacuole. Special organelles called lysosomes then fuse with the vacuole releasing digestive enzymes to digest the antigen.
- Natural killer cells are white blood cells that destroy abnormal cells by bringing about apoptosis.
- Phagocytes also have a role in specific defence, by presenting foreign antigens to lymphocytes.
- Specific lymphocytes are produced in response to specific antigens.
- B lymphocytes produce specific antibodies to combat specific foreign antigens.
- T lymphocytes target infected or damaged cells and bring about apoptosis.
- Lymphocytes are amplified through mitosis. This is called clonal selection. Some clones are used in combating the attack. Others become immunological memory cells.
- Endoparasites are able to mimic host antigens and avoid destruction.
- Many parasites show vast antigenic variation. This means being immune to one strain does not confer immunity to other strains.
- Some parasites are difficult to culture in the laboratory.
- Rapid parasite antigen change makes designing vaccines extremely difficult.
- The metabolisms of the host and parasite are often very similar, thus making it difficult to develop drugs that target the parasite without harming the host.

> **Summary continued**
> - Controlling parasites is easier where civil engineering has resulted in effective sanitation.
> - Controlling parasites is difficult in areas of overcrowding, tropical climates, developing countries, refugee camps or areas hit by natural disasters.
> - Where parasite control is improving, child mortality is falling. Children have better growth, development and the chance to increase their intelligence.

5.8 Extended response questions

The activity which follows presents an extended response question similar to the style that you will encounter in the examination.

You should have a good understanding of parasite niche, immune responses or the treatment and control of parasites before attempting these questions.

You should give your completed answer to your teacher or tutor for marking, or try to mark it yourself using the suggested marking scheme.

Extended response question: Parasite niche

Discuss the concept of the parasite niche. *(10 marks)*

Extended response question: Immune responses

Describe how mammals use immune responses to reduce the effects of parasites. *(10 marks)*

Extended response question: The treatment and control of parasites

Discuss the statement that 'there are major challenges in the treatment and control of parasites'. *(10 marks)*

5.9 End of topic test

End of Topic 5 test   Go online

Parasites and hosts form a symbiotic relationship.

Q22: This relationship is of _____ to the parasite. *(1 mark)*

a) benefit
b) detriment

....

Q23: This relationship is of _____ to the host. *(1 mark)*

a) benefit
b) detriment

....

Q24: Parasites have a _____ niche. *(1 mark)*

a) narrow
b) wide

....

Q25: This is due to _____ host specificity. *(1 mark)*

a) high
b) low

....

Q26: Tapeworms tend to lack a digestive system. What word that describes tapeworms and many parasites could be used here? *(1 mark)*

....

Q27: Liver flukes that reach sexual maturity in the livers of goats carry out asexual reproduction in snails as part of their life cycle. What term is used to describe the snail? *(1 mark)*

....

Q28: What term is used to describe the goat in the context of the previous question? *(1 mark)*

....

Q29: *Obeliscoides cuniculi* and *Trichuris leporis* are two species of tapeworm that parasitise rabbit intestines. Their niches are too similar and *Trichuris leporis* becomes locally extinct. What term describes this? *(1 mark)*

....

Q30: *Passalurus ambiguus* and *Dirofilaria scapiceps* are two other species of tapeworm that parasitise rabbit intestines. They are able to co-exist. What term describes this? *(1 mark)*

....

Q31: Which of the following statements that refer to nematode parasites are examples of parasites in their realised niche? Choose all that apply. *(1 mark)*

A) The nematode *Ascaris* inhabits the full length of the human intestine as it is not under any competition from other nematode species.

B) The nematode *Acsaris* inhabits the first 20% of the human intestine and the nematode *Strongyloides* inhabits the last 20% of the human intestine.

C) The mouse parasites *Heligmosomoides polygyrus*, found in the stomach, and *Litomosoides sigmodontis*, found in the small intestine, have adapted to suit different pH environments, thus reducing competition for resources.

....

Q32: Which of the following statements relating to parasites best describes the extended phenotype? *(1 mark)*

a) Degenerate
b) Modification of host behaviour
c) Transmitted by vectors
d) Virulence

....

Q33: Name two human diseases caused by parasites that require a second host. *(2 marks)*

....

Q34: Tuberculosis is a human disease that only requires a single host. What group of parasites is it caused by? *(1 mark)*

....

Q35: HIV and influenza are human diseases. What group of parasites are they caused by? *(1 mark)*

....

Q36: The size range of viruses is 20-400 __. *(1 mark)*

a) nm
b) μm
c) mm
d) cm

....

Q37: Which of the following can be components of a virus? *(3 marks)*

1. Internal carbohydrates
2. Lipid coat
3. Nucleus
4. Protein coat
5. RNA

....

Q38: Which of the following statements are true of the lipid membrane surround found in some viruses? *(2 marks)*

1. They are composed of the virus's materials.
2. They are composed of the host's materials.
3. They help the host detect the virus.
4. They help the virus remain undetected.

....

Q39: What is the name of the enzyme used by retroviruses to synthesis DNA from their RNA? *(1 mark)*

....

Q40: Arrange the steps of the replication of a virus in the correct order. *(1 mark)*

A) Virus uses the host cell's machinery and raw materials to replicate its DNA and synthesise protein coat.
B) Virus injects its DNA into the host cell.
C) These then leave the cell to infect new cells and the host cell undergoes lysis, bursting.
D) Virus interrupts the host cell's own metabolism, often entering the host's genome.
E) Virus attaches to the surface of the host cell.
F) The new DNA then enters the newly formed protein coats, thus producing many new viruses.

....

Q41: Complete the table using the statements listed. *(4 marks)*

Non-specific defence	Specific defence

Statement list: Action by B cells, Action by T cells, Clonal selection, Inflammatory response, Mucus, Phagocytes present antigens, Phagocytosis, Skin.

....

Q42: During phagocytosis, special organelles release digestive _____ into the vacuole containing the parasite. *(1 mark)*

....

Q43: What are the special organelles from the previous question known as? *(1 mark)*

....

Q44: In response to parasitic attack, lymphocytes are amplified through mitosis. What is this cellular response known as? *(1 mark)*

....

Q45: The role of lymphocytes in immune response is in: *(1 mark)*

a) phagocytosis.
b) clonal selection.
c) production of specific antibodies.
d) apoptosis of specific damaged cells.

....

Q46: Which of the following best describes how endoparasites evade destruction? *(1 mark)*

a) Mimics host antigens.
b) Mimics other parasites.
c) Kills lymphocytes.
d) Kills phagocytes.

....

Q47: Due to antigenic variation, last season's flu vaccine _____ be effective against this season's strain. *(1 mark)*

a) will
b) will not

....

Q48: NHS Scotland plans to vaccinate preschool children against mumps, measles and rubella using the MMR vaccination. Some parents are reluctant to vaccinate, being aware that if enough other children get vaccinated, their child will receive some protection anyway.

What is the ability to receive some protection even without having been vaccinated known as? *(1 mark)*

....

Q49: What is the study of outbreaks and spread of disease called? *(1 mark)*

....

Q50: Overcoming parasites can be _____. *(1 mark)*

a) easy
b) difficult

Q51: This is because the parasite evolution rate can be _____. *(1 mark)*

a) fast
b) slow

Q52: Another factor is that parasites are _____ to culture in the laboratory. *(1 mark)*

a) easy
b) difficult

Q53: Additionally, the host and parasite have very _____ metabolisms. *(1 mark)*

a) different
b) similar

Q54: What is the main reason that parasite abundance in the UK is so low? *(1 mark)*

Q55: Which condition promotes the spread of parasites as a result of overcrowded refugee camps and a lack of good sanitation? *(1 mark)*

a) Natural disasters
b) Tropical climates
c) War

Q56: Which condition promotes the spread of parasites due to a large population of vectors? *(1 mark)*

a) Natural disasters
b) Tropical climates
c) War

Q57: Which condition promotes the spread of parasites by limiting the supply of fresh drinking water or damaging sewage works? *(1 mark)*

a) Natural disasters
b) Tropical climates
c) War

Q58: Parasites are more abundant in _____ countries. *(1 mark)*

a) developed
b) developing

Q59: If parasite levels fall, infant mortality _____. *(1 mark)*

a) decreases
b) increases

Q60: If parasite levels fall, children will see _____ growth and development. *(1 mark)*

a) decreased
b) increased

Unit 2 Topic 6

Organisms and evolution test

Organisms and evolution test

Go online

Field techniques for biologists

Q1: What kind of sampling could involve placing a quadrat every 3 metres along a sandy shore? *(1 mark)*

....

Q2: Perwinkles are suitable for mark and recapture experiments. The formula

$$N = \frac{MC}{R}$$

is used to estimate periwinkle population size following mark and recapture.

- N = population estimate;
- M = number first captured, marked and released;
- C = total number in second capture;
- R = number marked in second capture.

Periwinkles were marked and recaptured on a rocky shore near Creetown in Galloway. The results are shown.

- Number of periwinkles first captured, marked and released = 100
- Number of marked periwinkles in second capture = 50
- Number of unmarked periwinkles in second capture = 90

Calculate the estimated population of periwinkles on the rocky shore near Creetown. *(1 mark)*

....

Q3: What is the study of animal behaviour called? *(1 mark)*

....

Q4: An ethogram is a list of all the behaviours shown by species. This can be used to produce time budgets for wild species. Measurements taken to produce a time budget include latency, and the _____ and _____ of particular behaviours. *(1 mark)*

....

Q5: When studying animal behaviour, anthropomorphism must be avoided. Which of the following notes taken by an ethologist while observing a group of young wolf cubs show anthropomorphism? Choose two. *(2 marks)*

A) Wagging their tails.
B) Having fun.
C) Making friends.
D) Barking.

Evolution

Q6: Over several generations in a small population the gene pool seemed to change randomly. What process is likely to be responsible? *(1 mark)*

....

Q7: What causes variation in traits to arise in populations? *(1 mark)*

....

Q8: What can mutations be aside from being either beneficial or harmful? *(1 mark)*

....

Q9: In evolutionary theory, fitness can be termed absolute or relative. Which of the following is the correct definition of absolute fitness? *(1 mark)*

a) Frequencies of a particular genotype from one generation to the next.
b) Frequencies of a particular phenotype from one generation to the next.
c) Surviving offspring of one genotype compared with other genotypes.
d) Surviving offspring of one phenotype compared with other phenotypes.

....

Q10: In evolutionary theory, fitness can be termed absolute or relative. Which of the following is the correct definitions of relative fitness? *(1 mark)*

a) Frequencies of a particular genotype from one generation to the next.
b) Frequencies of a particular phenotype from one generation to the next.
c) Surviving offspring of one genotype compared with other genotypes.
d) Surviving offspring of one phenotype compared with other phenotypes.

....

Q11: The rate of evolution can be increased by several factors. Which of the following will increase the rate of evolution? Choose three. *(3 marks)*

A) Cold environments.
B) Horizontal gene transfer.
C) Long generation times.
D) Short generations times.
E) Warm environments.

....

Q12: A change in the traits of one species acting as a selection pressure on another species with which it frequently interacts is called _____. *(1 mark)*

....

Q13: After many generations, tapeworm parasite species evolved hooks to be able to grip onto their human host's intestinal lining. After many subsequent generations, humans evolved to produce an enzyme capable of dislodging the tapeworm's hooks.

What name is given to this evolutionary arms race? *(1 mark)*

Variation and sexual reproduction

Q14: Which of the following are true of sexual reproduction? Choose two. *(2 marks)*

A) Horizontal gene transfer in bacteria is an example.
B) Great variation may occur throughout offspring.
C) No variation in offspring.
D) Only half the population is able to produce offspring.

....

Q15: Which of the following is true of asexual reproduction? *(1 mark)*

a) Driving force for the Red Queen's arms race.
b) Great variation in offspring.
c) Only half of the population is able to produce offspring.
d) Vegetative propagation in plants is an example.

....

Q16: In an environment where female komodo dragons struggle to find a male to mate with, what method of asexual reproduction do they use? *(1 mark)*

....

Q17: Which of the following statements about homologous chromosomes is false? *(1 mark)*

a) Each inherited from a different parent.
b) Same centromere position.
c) Same genes at different gene loci.
d) Same size.

....

Q18: When does crossing over occur? *(1 mark)*

a) Meiosis I
b) Meiosis II

....

Q19: When does independent assortment occur? *(1 mark)*

a) Meiosis I
b) Meiosis II

....

Q20: How many cells does meiosis produce? *(1 mark)*

a) 2
b) 4

....

Q21: What kind of cells are produced by meiosis? *(1 mark)*

a) Diploid
b) Haploid

....

Q22: The cells that are produced by meiosis are genetically _____. *(1 mark)*

a) different
b) identical

....

Q23: Male fruit flies have an X and a smaller Y chromosome. What term describes this condition? *(1 mark)*

....

Q24: Genes J, K, L, M and N are all found on the same chromosome. What term is used to describe genes located on the same chromosome? *(1 mark)*

....

Q25: The recombination frequencies of genes J, K, L, M and N are given in the following table.

Genes	Recombination frequency (%)
N and J	16
M and L	6
K and M	27
K and L	33
K and N	4
J and M	7

In which order would these genes appear on the chromosome? *(1 mark)*

....

Q26: The following diagram shows two homologous chromosomes.

| | A | b | c | D |
| | A | b | c | D |

| | a | B | C | d |
| | a | B | C | d |

Between which of the following alleles would the frequency of recombination be greatest? *(1 mark)*

a) A and B
b) B and D
c) A and D
d) B and C
e) A and C

....

Q27: Many species of snail have both male and female reproductive systems. What term is used to describe this? *(1 mark)*

....

Q28: In some species, the sex of offspring is not controlled by genetic factors. What factors control it in these cases? *(1 mark)*

....

Q29: In fruit flies, males are heterogametic and eye colour is carried on the X chromosome. The allele for red eyes R is dominant to the allele for white eyes r. If the mother had white eyes and the father had red eyes, what will be the appearance of the offspring? *(1 mark)*

a) White eyed females and red eyed males.
b) White eyed females and white eyed males.
c) Red eyed females and white eyed males.
d) Red eyed females and red eyed males.

....

Q30: Red green colour blindness is a recessive human sex-linked trait. An X chromosome in every cell gets inactivated. Which of the following statements explain why the female does not then show red green colour blindness where the normal allele has been inactivated? Choose two. *(2 marks)*

A) Inactivation is random.
B) At least half of her cells will have a working copy of the gene.
C) Only the chromosome carrying the red green colour blindness allele gets inactivated.

Sex and behaviour

Q31: Which of the following correctly identifies the characteristics of sperm and eggs? *(1 mark)*

a) Sperm: many; Eggs: many.
b) Sperm: few; Eggs: contain an energy store.
c) Sperm: contain an energy store; Eggs: few.
d) Sperm: many; Eggs: contain an energy store.

....

Q32: Which sex is considered to make a greater sexual investment? *(1 mark)*

....

Q33: Which of the following statements about parental investment is true? *(1 mark)*

a) In a stable environment, organisms produce many, cheap offspring.
b) r-selected organisms produce many offspring, but provide no parental care.
c) All organisms can be easily classified as either r-selected or K-selected.
d) Organisms inhabiting stable environments will likely have one large reproductive event.

....

Q34: Birds of paradise show distinct differences between males and females. What term is used to describe the differences between males and females? *(1 mark)*

....

Q35: Which of the following potentially increases a male's access to females? Choose three. *(3 marks)*

A) Hunting.
B) Large size.
C) Sneaking.
D) Use of weaponry.

....

Q36: What is the irreversible developmental process that occurs during a critical time period in young birds and may influence mate choice later in life called? *(1 mark)*

....

Q37: What refers to smaller, duller females that tend to camouflage better than males? *(1 mark)*

a) Fitness
b) Inconspicuous
c) Lekking
d) Reversed sexual dimorphism

....

Q38: What refers to females tending to be larger and more conspicuous than males? *(1 mark)*

a) Fitness
b) Inconspicuous
c) Lekking
d) Reversed sexual dimorphism

....

Q39: What is a behaviour shown by male birds and insects, where they display to females in a communal area? *(1 mark)*

a) Fitness
b) Inconspicuous
c) Lekking
d) Reversed sexual dimorphism

....

Q40: What can be assessed by females in terms of good genes and low parasite burden? *(1 mark)*

a) Fitness
b) Inconspicuous
c) Lekking
d) Reversed sexual dimorphism

Parasitism: Niche

Q41: Which of the following statements are true of parasites? Choose three. *(3 marks)*

A) Wide niche.
B) Narrow niche.
C) High host specificity.
D) Low host specificity.
E) Often lack certain structures and organs.
F) Have a full complement of structures and organs.

....

Q42: Nematodes in the intestines of turtles lack a digestive system and are said to be: *(1 mark)*

a) competitive exclusion.
b) degenerate.
c) fundamental.
d) realised.
e) resource partitioning.

....

Q43: When two species have such similar niches that one becomes locally extinct as a result in interspecific competition, this is known as: *(1 mark)*

a) competitive exclusion
b) degenerate
c) fundamental
d) realised
e) resource partitioning

....

Q44: A species occupies its _____ niche when interspecific competition is absent. *(1 mark)*

a) competitive exclusion
b) degenerate
c) fundamental
d) realised
e) resource partitioning

....

Q45: A species occupies its _____ niche when interspecific competition is present. *(1 mark)*

a) competitive exclusion
b) degenerate
c) fundamental
d) realised
e) resource partitioning

....

Q46: Species participating in interspecific completion may co-exist by: *(1 mark)*

a) competitive exclusion.
b) degenerate.
c) fundamental.
d) realised.
e) resource partitioning.

Parasitism: Parasitic Life cycles

Q47: What name is given to the host in or on which the parasite reaches sexual maturity? *(1 mark)*

....

Q48: Another host is sometimes needed as a vector or for the parasite to complete its life cycle. What name is given to this host? *(1 mark)*

....

Q49: A parasite has the potential to harm its host. What term refers to this? *(1 mark)*

....

Q50: Which statement would be referred to as part of the extended phenotype of a parasite? *(1 mark)*

a) Alteration of host foraging.
b) Similarity to the host cell metabolism.
c) Transmission
d) Virulence

....

Q51: Schistosomiasis is a human disease caused by Schistosoma: *(1 mark)*

a) flatworms.
b) roundworms.

....

TOPIC 6. ORGANISMS AND EVOLUTION TEST

Q52: Another human disease caused by a parasite is malaria. How many hosts does the *Plasmodium* parasite require to complete its life cycle? *(1 mark)*

a) 1
b) 2

....

Q53: Influenza and HIV are both human diseases caused by: *(1 mark)*

a) bacteria.
b) viruses.

....

Q54: Ectoparasitic arthropods, such as ticks, complete their life cycle on one host. Which of the following statements are true of arthropods? Choose three. *(3 marks)*

A) Jointed legs.
B) Contain a shell.
C) Body not divided into segments.
D) Segmented body.
E) Exoskeleton.

....

Q55: Which of the following statements are true of virus structure? Choose three. *(3 marks)*

A) They are surrounded by a protein coat.
B) They contain only DNA.
C) They contain RNA or DNA.
D) They contain only RNA.
E) Some have a lipid membrane surround.
F) All have a lipid membrane surround.

....

Q56: Arrange the steps of the replication of a virus in the correct order. *(1 mark)*

A) Virus uses the host cell's machinery and raw materials to replicate its DNA and synthesise protein coat.
B) Virus injects its DNA into the host cell.
C) These then leave the cell to infect new cells and the host cell undergoes lysis, bursting.
D) Virus interrupts the host cell's own metabolism, often entering the host's genome.
E) Virus attaches to the surface of the host cell.
F) The new DNA then enters the newly formed protein coats, thus producing many new viruses.

....

Q57: Retroviruses contain RNA rather than DNA. On inserting the RNA into the host, it must first be converted into DNA. What enzyme catalyses this reaction? *(1 mark)*

Parasitism: Transmission and virulence

Q58: Choose the correct words to complete the following sentence.

The lower/higher the transmission rate, the greater the virulence.

Parasitism: Defence against parasitic attack

Q59: When the skin becomes wounded, the surrounding area becomes red and warm due to increased blood flow. What name is given to this response? *(1 mark)*

....

Q60: Phagocytosis is a non-specific cellular defence. Arrange the stages in the process of phagocytosis in the correct order. *(1 mark)*

A) Lysosomes fuse with the vacuole.
B) Pathogen is digested by enzymes.
C) Phagocyte engulfs the pathogen into a vacuole.
D) Product of pathogen digestion released into host cell cytoplasm.

....

Q61: Lymphocytes respond to foreign _____ by producing special proteins. *(1 mark)*

a) antibodies
b) antigens
c) apoptosis
d) clonal selection

....

Q62: The special proteins are called: *(1 mark)*

a) antibodies.
b) antigens.
c) apoptosis.
d) clonal selection.

....

Q63: Lymphocytes destroy infected cells by bringing about: *(1 mark)*

a) antibodies.
b) antigens.
c) apoptosis.
d) clonal selection.

....

TOPIC 6. ORGANISMS AND EVOLUTION TEST

Q64: During specific responses, lymphocytes become amplified through mitosis. This process is called: *(1 mark)*

a) antibodies.
b) antigens.
c) apoptosis.
d) clonal selection.

....

Q65: It is important to study the outbreak and spread of infectious diseases. What name is given to this field of biology? *(1 mark)*

....

Q66: NHS Scotland advises that all pre-school and primary school children receive the influenza vaccine in order to reduce the spread of influenza, particularly to vulnerable older adults, e.g. grandparents. What is this protection method an example of? *(1 mark)*

Parasitism: Defence against parasitic attack

Q67: Parasites often _____ the host's immune system. *(1 mark)*

a) promote
b) suppress

....

Q68: Parasites often _____ host size. *(1 mark)*

a) decrease
b) increase

....

Q69: Parasites often _____ host reproduction. *(1 mark)*

a) decrease
b) increase

Parasitism: Challenges in treatment and control

Q70: Developing countries, such as Malawi, find it more difficult to control parasites than developed countries, such as the UK. Which of the following statements does not explain why this is the case? *(1 mark)*

a) Developing countries usually have tropical climates.
b) Developing countries may have more natural disasters.
c) Developing countries have poor sanitation.
d) Developing countries have higher populations.
e) There are more vectors in developing countries, e.g. mosquitoes.
f) Overcrowding is more common in developing countries.
g) Many developing countries are in conflict.

Investigative Biology

1 Scientific principles and process — 245
- 1.1 Scientific method — 247
- 1.2 Scientific communication and literature — 250
- 1.3 Scientific ethics — 252
- 1.4 Learning points — 254
- 1.5 End of topic test — 255

2 Experimentation — 257
- 2.1 Key terms — 260
- 2.2 Pilot study — 261
- 2.3 Experimental design — 263
- 2.4 Sampling — 270
- 2.5 Reliability — 272
- 2.6 Presentation of data — 274
- 2.7 Extended response question — 278
- 2.8 Learning points — 279
- 2.9 End of topic test — 282

3 Critical evaluation of biological research — 285
- 3.1 Evaluating background information — 287
- 3.2 Evaluating experimental design — 288
- 3.3 Evaluating data analysis — 289
- 3.4 Evaluating conclusions — 293
- 3.5 Learning points — 296
- 3.6 End of topic test — 298

4 Investigative biology test — 299

Unit 3 Topic 1

Scientific principles and process

Contents

1.1 Scientific method 247
      1.1.1 Scientific cycle model 247
      1.1.2 Null hypothesis and independent verification 249
1.2 Scientific communication and literature 250
1.3 Scientific ethics 252
1.4 Learning points 254
1.5 End of topic test 255

Prerequisites

You should already know that:

- scientific reports are written in standard format;
- experiments should be repeated for reliability;
- experiments should cause no harm or distress to living things or the environment;
- risk assessments should be written and followed when carrying out experiments.

> **Learning objective**
>
> By the end of this topic, you should be able to:
>
> - describe the scientific cycle, naming the four key parts;
> - state that in science, ideas, theories and hypotheses are constantly being refined;
> - define the term null hypothesis;
> - explain that independent verification is necessary when accepting new scientific concepts;
> - state that negative results can be important, using medical advances as an example;
> - state that experiments and findings must be communicated in a standard format that allows repetition by other scientists;
> - state that experimentation and findings will be subject to peer review and critical evaluation;
> - explain that where the wider media will provide further critical evaluation, and be able to identify where such critical evaluation is inaccurate or biased;
> - explain the importance of scientific ethics in terms of unbiased reporting, reliable references and avoiding plagiarism;
> - state that animals should only be used when absolutely necessary and any harm to them minimal;
> - explain the ethics involved when working with humans, in terms of consent and the right to withdraw from the study at any time;
> - describe the influence of risk assessments, legislation, regulation, policy and funding in scientific research.

TOPIC 1. SCIENTIFIC PRINCIPLES AND PROCESS

1.1 Scientific method

Science is the gathering and organisation of testable and reproducible knowledge. In the scientific cycle there are four key parts:

- hypothesising, questioning and debating a particular idea or area of science;
- investigating through observing, researching or experimenting;
- analysis of data and results from experiments through comparing, interpreting or applying statistics;
- evaluation of results and conclusions are formed.

Based on analysis and evaluations, a new or refined hypothesis may be offered and so the cycle continues (see the following diagram).

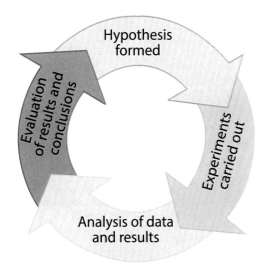

The scientific cycle

1.1.1 Scientific cycle model

In science, ideas, models and theories are constantly being refined. At any one time the current theory is considered to be the best explanation or answer to a hypothesis. After further experimentation, analysis and evaluation an updated theory or model may be offered, thus superseding the previous theory.

Reaching the current fluid mosaic model of the plasma membrane illustrates the scientific cycle perfectly.

The evolution of plasma membrane models

Work on the plasma membrane theory started in 1895, when Charles Overton proposed that lipids were the main components of membranes. He had observed lipid soluble substances enter cells much more quickly than lipid insoluble substances. Of course in 1895, it was still 50 years prior to membranes being viewed with an electron microscope. Twenty years later the idea of proteins also being part of the membrane was introduced.

In 1917, Irving Langmuir introduced the idea that the lipids may in fact be phospholipids, with hydrophilic heads during his work with artificial membranes. Eight years later, two Dutch scientists, E. Gorter and F. Grendel, put forward the hypothesis that cell membranes were phospholipid bilayers, thus explaining how the hydrophobic tails could avoid contact with water. Gorter and Grendel carried out measurements using red blood cells, finding that the phospholipid content to be double that of the surface area of the cells, permitting a double layer. It was discovered later that these two Dutch scientists had miscalculated both the surface area of the cell and the phospholipid quantity, however, the two mistakes cancelled each other out. Their conclusion was correct, but based on erroneous data.

In 1935, Hugh Davson and James Danielli started hypothesising the arrangement of the proteins. They postulated that the phospholipid bilayer was sandwiched between two layers of globular protein. This model seemed to be supported with the introduction of electron microscopy to the study cells in the 1950s. By the 1960s, the Davson-Daneilli sandwich theory was the favoured plasma membrane and organelle membrane model. However, by the end of the 1960s, many cell biologists were noticing two issues with the model:

1. First, not all membranes were identical under the electron microscope. The plasma membrane is 7-8 nm thick with a three-layered structure. Conversely, the mitochondrial inner membrane measures only 6 nm in thickness. The protein content of mitochondrial membranes is greater than that of plasma membranes. Furthermore, the type of phospholipids present was found to differ between different membranes.

2. The second issue with the sandwich model was explaining the protein arrangement. In the sandwich model, the proteins were hypothesised to provide a layer on the membrane surface. This would pose a problem for the sections of the proteins that are hydrophobic. In 1972, a new hypothesis solving this problem was proposed by S. J. Singer and G. Nicolson. Their model saw the proteins scattered and inserted into the phospholipid bilayer. This is the theory we currently accept, the **fluid mosaic model**.

As can be seen, the theory of the plasma membrane was nearly 80 years in the making. This is not unusual in science. Other examples of the scientific cycle are:

- research on DNA structure;
- discovery of plant hormones;
- evidence that genes specify proteins;
- Gregor Mendel's approach to genetics;
- research to identify model organisms for scientific study.

Scientific cycle	Go online

Q1: Put the following steps in the scientific cycle into the correct order:

- Analysing data through comparing, interpreting and applying statistics
- Debating ideas and coming up a hypothesis to test
- Designing appropriate experiments
- Evaluating results
- Forming conclusions
- Observing and collecting data from experiments
- Refining the original hypothesis
- Researching others' work

1.1.2 Null hypothesis and independent verification

Can negative results be positive?

Failure to find an effect (i.e. a negative result) is a valid finding, as long as an experiment is well designed. Even conflicting data or conclusions can be resolved through careful analysis and evaluation. Where a resolution cannot be reached, additional and often more creative experimentation can be carried out. In medicine and pharmaceuticals, reporting of negative results is essential in the pursuit of cures for degenerative and acute diseases.

The null hypothesis

The null hypothesis is a hypothesis based on the default position, thus that two variables are unrelated. An example would be 'all daisies have the same number of petals'. Clearly this theory can be easily disproved through discovery of two daisies with a different number of petals. Often scientists will state a null hypothesis and set out to falsify it.

Hypothesis: Larger males are more likely to be selected by a mate.

Null hypothesis: Larger males are not more likely to be selected by a mate.

Should a relationship between larger males and selection be observed through experimentation, the null hypothesis could easily be rejected. Rather than setting out to prove hypotheses, scientists often set out to disprove or falsify null hypotheses.

Independent verification

Scientific ideas only become accepted once they have been verified (or alternatives falsified) independently; one-off results are treated with caution, which is why it is essential to do complete independent repeats of an experiment. Repetition should be done at a different time with a completely new set of ingredients, organisms and chemicals.

An investigation into whether athletes' body clocks affect competition performance Go online

In this investigation male and female athletes from different sports were used to ascertain whether body clock affected competition performance. It has long been known that some individuals are naturally early risers and others more night owls.

Different athletes train at different times of the day, with a number of factors from work and family life to accessing equipment being amongst the influencing factors. Athletes have little say over timing of competitions though, with these being governed by facilities and often television rights. Early rising female lacrosse players have been shown to have a peak performance at noon, whereas late risers didn't peak until around 8.00 p.m. (Parker et al, 2009).

Obviously in a team sport, it is difficult to suit all players. Clearly, if the findings of the female lacrosse players are found in other sports too, and in both males and females, then this could be highly significant. In some sports, such as sprinting, the margin between a medal or not could come down to hundredths of a second.

Q2: What was the aim of this investigation?

....

Q3: What was the null hypothesis of this investigation?

....

Q4: If the result is a negative result, who might benefit from this being reported and why?

1.2 Scientific communication and literature

Scientific reports follow a basic format:

- title;
- aims and hypotheses;
- methods;
- results;
- analysis of results;
- discussion and evaluation.

Scientific reports should be written in a manner that allows other scientists to repeat the experiment for verification or further work.

Scientific findings may be communicated via a range of different means. Scientists from a range of fields may share a selection of findings at conferences and seminars via lectures, talks and poster displays. Furthermore, where findings have been subject to peer review, they may be reported via scientific journals, thus following the common scientific report format.

Review articles

Where a number of scientific findings have been made in a particular field, most scientific journals will publish a review article which sums up the work and findings carried out. These review articles are usually written by experts who are well respected experts in their chosen area of scientific research.

Referencing

When referencing:

- cite and reference these in a standard format;
- use only references that are reliable and unbiased;
- care must be taken to avoid **plagiarism**.

Plagiarism can be avoided by putting statements into your own words and clearly citing sources. Never include large chunks of copied material or long quotations.

Referencing - standard format

- **Books**

    Author(s) (surname followed by initials). (Year). Book title. Place: Publisher.

    Example reference: Raven, P.H., Evert, R. F. and Eichhorn, S. E. (1999). Biology of Plants (6ed). New York: W. H. Freeman and Company.

- **Journal articles**

    Author(s) (surname followed by initials). (Year). Article title. Journal Title Volume (issue), pages.

    Example reference: Tanner, K. (2012). Promoting Student Metacognition. Life Science Education 11, 113-120.

- **Web page**

    Author. (Year). Title. Available at: web address of document [Accessed: Day Month Year].

    Example reference: Robins, C. (2015). Athletes Vs Non-Athletes Heart Rate. Available at: http://www.livestrong.com/article/82658-athletes-vs.-nonathletes-heart-rate/ [Accessed: 21/07/16]

If there is no clear author, the organisation should replace the author.

If there is no date, state (no date) instead.

1.3 Scientific ethics

Integrity and honesty

All findings must be presented in an unbiased manner. Other scientists may of course dispute or disagree with findings; however, trust in findings is increased through referencing others' findings.

Peer review

Scientists who specialise in a particular field read reports prior to publication, assessing reliability. These specialists may advise the writer to make changes to enhance the scientific rigour of the piece before it can be published. Once the article is finally published, scrutiny of the findings continues through further critical evaluation via coverage in the wider media. It is vital that this scientific information is presented to the wider media in an unbiased and understandable manner.

Increasing the public understanding of science and the issue of misrepresentation of science in the media can be difficult. Sometimes members of the public do not feel they can access the information due to the technical terms and jargon used. With current topical science surrounding stem cell research, genetic engineering, DNA technology, food security and pharmacology it is even more important to present the information in an unbiased and accessible manner to allow the public to make well-formed opinions.

Too often, the media has created panic regarding scientific matters that have adversely affected individuals. One example is that of the controversy surrounding the MMR vaccination in the late 1990s. Parents were put in a very difficult position about whether to have their children vaccinated against these serious diseases when an article was published in 1998 by Andrew Wakefield based on a very small sample size, with ambiguous results suggesting that there might be a link between MMR and autism. Wakefield's publications included a review paper with no new evidence, published in a minor journal, and two papers on laboratory work where he provided data that the measles virus had been found in tissue samples taken from children who had autism and bowel problems.

Wide media coverage followed, with very emotive stories from distressed parents. The health service came under attack and controversy grew further when the Prime Minister at the time refused to say whether his young son had been immunized or not. In very recent years, outbreaks of these diseases have been reported in unvaccinated teenagers. Mumps, as an example, is particularly dangerous to teenage boys who may suffer infertility as a result of the disease. Although this study has now been completely discredited, some of the general public still believe these findings. This is a clear example of the consequences of poor science that has caused significant harm to children.

Replication of experiments by peers also reduces the opportunity for dishonesty or any deliberate misuse of science.

Animal studies

Using animals in studies is often unavoidable, but does lead to advances in medicine. Where the use of animals is unavoidable, the 3Rs are recommended:

1. *Replacement* - can the animal be replaced with another? Who decides which animals are acceptable for use and which are not though?
2. *Reduction* - can fewer animals be used?
3. *Refinement* - can the procedure be refined to reduce human error?

TOPIC 1. SCIENTIFIC PRINCIPLES AND PROCESS

It is essential that scientists use models or alternatives to animals where possible. However, should using an animal be unavoidable this must be well documented. Any experiments involving animals are strictly controlled, requiring licensed premises and licenses to carry out each piece of research. Researchers must provide full justification of which animals are to be used and why. They must also state the expected positive impact on humans and society that using these animals will bring.

Human studies

In any studies involving humans, informed consent should always be sought. If participants are under 16, a parent or carer must also provide informed consent. Participants must be aware of the right to withdraw data at any time during the investigation. Participant confidentiality must be maintained too.

Through providing informed consent, participants will be able to assess if their participation may contribute to findings that could impact on society or the environment.

Justification

All scientific investigations must be justifiable, thus clearly have benefits to society or the environment, or merely the pursuit of scientific knowledge.

Scientific investigations are influenced by:

1. risk assessments that must take account of that safety of humans or animals involved and any impact on the environment;
2. regulation, policy and licensing by governments which aim to limit the potential for the misuse of studies and data;
3. funding which can influence the direction and pace of scientific progress.

Scientific ethics   Go online

Q5: Match up the terms about using animals in scientific studies with their definitions.

Term	Definition
	Ensuring competence in the experimental technique to reduce human error.
	Using a different type of animal in the study.
	Using fewer animals in the study.

Terms: Reduce, Refine, Replace.

An investigation into whether athletes' body clocks affect competition performance

It has long been known that some individuals are naturally early risers and others more night owls. Different athletes train at different times of the day, with a number of factors from work and family life to accessing equipment being amongst the influencing factors. Athletes have little say over timing of competitions though, with these being governed by facilities and often television rights. Early rising female lacrosse players have been shown to have a peak performance at noon, but late risers didn't peak until around 8.00 p.m. (Parker et al, 2009).

Obviously in a team sport, it is difficult to suit all players. Clearly, though, these findings, if shown in other events and in both males and females, could be highly significant. In some sports, such as sprinting, the margin between a medal or not could come down to hundredths of a second.

A professor of Sports Science wanted to further research whether an athlete's body clock had an impact on what time of day they would perform better in a competition.

Three groups of 20 athletes (10 male and 10 female) from a range of sports were used:

- Group 1 - athletes that were 'morning people' (early morning risers);
- Group 2 - athletes that were 'night owls' (late morning risers);
- Group 3 - athletes that had no preference to getting up late or early.

Q6: Describe any ethical considerations that should have been part of the design of this investigation.

....

Q7: In the background information, the author states: 'Early rising female lacrosse players have been shown to have a peak performance at noon, but late risers didn't peak until around 8.00 p.m.'

How is this statement justified by the author?

1.4 Learning points

Summary

- Scientific cycle involves setting testable hypotheses or null hypotheses, followed by suitable experimentation and subsequent data collection. Data should then be analysed and results evaluated. Conclusions can then be drawn and hypotheses modified accordingly. These modified hypotheses can then be tested and so the cycle continues.
- Null hypotheses are hypotheses stating that one variable will have no effect on the other.
- Scientific ideas require independent verification before being published and supported

TOPIC 1. SCIENTIFIC PRINCIPLES AND PROCESS

> **Summary continued**
>
> by similar findings in a number of other investigations before being accepted.
>
> - Scientific findings and investigations must be reported in the standard format, with enough detail to be repeated by other scientists.
> - All findings are subject to critical evaluation from peers, in peer review and by the public through the wider media.
> - When reporting findings, this must be done in an unbiased manner, backed by reliable sources, referenced in the standard format.
> - Plagiarism must be avoided.
> - When using animals, harm should be minimised and scientists should consider replacement, reduction and refinement.
> - When using humans, consent is essential and subjects must be aware of their right to withdraw their data at any time.
> - Confidentiality must be provided to human subjects.
> - Scientific investigations must follow a working risk assessment.
> - Legislation, regulation, policy and funding can all influence scientific research.

1.5 End of topic test

End of Topic 1 test Go online

Q8: A scientist predicts that the speed of cycling will have no effect on the oxygen consumption of cyclists. What term is used to describe this statement? *(1 mark)*

....

Q9: Before publication of a scientific article, others scientists read the article and make any comments or suggestions to the author prior to publication. What is this process called? *(1 mark)*

....

Q10: Which of the following statements about results are true: *(4 marks)*

a) Results can be negative.
b) Results can be positive.
c) Results will be verified by a friend.
d) Results will be verified independently.
e) One-off results are easily accepted.
f) Results may cause a hypothesis to be rejected.

Q11: Complete the following sentence about scientific literature and communication.

Scientific reports should be written in a manner that allows other scientists to .... the experiment for verification or further work. *(1 mark)*

Q12: Referencing: *(1 mark)*

a) can reduce bias.
b) has a standard format.
c) is plagiarising others' work.

Q13: Identify the 3 'R's of using animals in scientific studies. *(1 mark)*

a) Reduce, reuse and replace
b) Reduce, replace and refine
c) Reuse, repeat and replace
d) Random, refine and reduce

Q14: Which of the following statements about using humans in studies are correct? *(2 marks)*

a) Consent is only required for children.
b) Consent is only required for adults.
c) Consent is required for all.
d) Subjects can withdraw data at any time.
e) Subjects can only withdraw data until the halfway point.
f) Subjects can never withdraw their data.

Unit 3 Topic 2

Experimentation

Contents

2.1 Key terms 260
2.2 Pilot study 261
    2.2.1 Pilot study examples 262
2.3 Experimental design 263
    2.3.1 Independent and dependent variables 263
    2.3.2 Confounding variables 265
    2.3.3 Controls 266
    2.3.4 In vitro vs. in vivo experiments 269
2.4 Sampling 270
2.5 Reliability 272
2.6 Presentation of data 274
2.7 Extended response question 278
2.8 Learning points 279
2.9 End of topic test 282

Prerequisites

You should already know that:

- in an experiment the independent variable is the variable which is manipulated or changed by the investigator and the dependent variable is the variable which is measured;

- within an experiment there are certain key variables which must be controlled in order to obtain a valid result;

- the results of control groups are used for comparison with treatment results;

- experiments are repeated and average results are calculated to improve the reliability of the results.

Learning objective

By the end of this topic, you should be able to:

- define the terms, validity, reliability, accuracy and precision;
- state that a pilot study is used to help plan procedures, assess validity and check techniques, this allows evaluation and modification of experimental design;
- explain the importance of a pilot study in determining an appropriate range of values for the independent variable;
- state that pilot studies allow the investigator to establish the number of repeat measurements required to give a representative value for each independent datum point;
- define the terms independent and dependent variables;
- state that experiments involve the manipulation of the independent variable by the investigator;
- state that the experimental treatment group is compared to a control group;
- describe the use and limitations of simple (one independent variable) and multifactorial (more than one independent variable) experimental designs;
- state that in some cases investigators may use groups that already exist, so there is no truly independent variable;
- describe the role of observational studies;
- state that due to the complexities of biological systems, other variables besides the independent variable may affect the dependent variable;
- state that these confounding variables must be held constant if possible, or at least monitored so that their effect on the results can be accounted for in the analysis;
- describe randomised block design;
- describe the use of negative and positive controls;
- describe the use of placebos and the placebo effect;
- define the terms in vitro and in vivo;
- describe the advantages and disadvantages of in vivo and in vitro studies;
- describe the role of sampling;
- state that the extent of the natural variation within a population determines the appropriate sample size and more variable populations require a larger sample size;
- describe the characteristics of a representative sample;
- describe random, systematic and stratified sampling;

> **Learning objective continued**
>
> - state that variation in experimental results may be due to the reliability of measurement methods and/or inherent variation in the specimens;
> - describe how the precision and accuracy of repeated measurements can be determined;
> - state that the natural variation in the biological material being used can be determined by measuring a sample of individuals from the population and the mean of these repeated measurements will give an indication of the true value being measured;
> - state that the range of values is a measure of the extent of variation in the results and if there is a narrow range then the variation is low;
> - describe the importance of independent replication;
> - state that discrete and continuous variables give rise to qualitative, quantitative, or ranked data;
> - state that the type of variable being investigated has consequences for any graphical display or statistical tests that may be used;
> - calculate mean, median and mode;
> - use box plots to show variation within and between data sets including median, lower quartile, upper quartile and inter-quartile range;
> - state that correlation exists if there is a relationship between two variables;
> - describe the relationship between correlation and causation;
> - describe positive and negative correlations;
> - describe the difference between a strong and a weak correlation.

2.1 Key terms

There are four key terms to consider when conducting an experiment and analysing results.

1. Validity
2. Reliability
3. Accuracy
4. Precision

Results are considered to be valid if confounding variables have been controlled so that any measured effect is likely to be due to the independent variable alone. Validity should be considered during the planning and set-up of an experiment. Any confounding variables which cannot be controlled should be closely monitored and taken into consideration when analysing results.

One off results from an experiment are inherently unreliable as they could has arisen by chance. Experiments should be repeated by both the initial researcher and an independent scientist; if the values generated between repeats and independent replicates are consistent then the results are considered to be reliable.

Accuracy refers to data, or means of data sets, that are close to the true value while precision refers to whether the measured values are close to each other. These are two terms which are often confused and used interchangeably despite their entirely different meanings. To consider the difference we can use the following darts analogy.

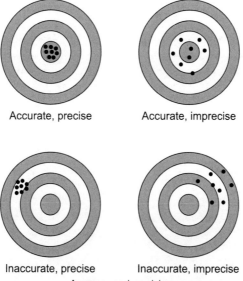

Accuracy and precision

Results are considered to be accurate if they are on or near the bullseye (in this analogy, the bullseye represents the 'true value'). Results are considered to be precise if all the darts land in close proximity to each other.

2.2 Pilot study

A pilot study is a small-scale investigation of a planned research project. The main aim of a pilot study is to assess validity and check procedures/techniques; this allows evaluation and modification of the experimental design. Pilot studies are also known as 'feasibility' studies since they investigate whether a suggested protocol is feasible, i.e. capable of delivering valid and reliable results.

Pilot studies are becoming integral to the development of any research project due to their ability to flag up potential issues with a main study and therefore prevent costly mistakes, time wasting or even failure of the project as a whole. Conducting a pilot study does not guarantee success of a project but a good pilot study increases the likelihood of success.

A pilot study can be used to develop and/or practice protocols in order to:

- ensure the experimental design is valid;
- ensure an appropriate range of values for the independent variable;
- identify a suitable number of replicates required to give a true value for each independent datum point.

The following diagram details the advantages of conducting a pilot study.

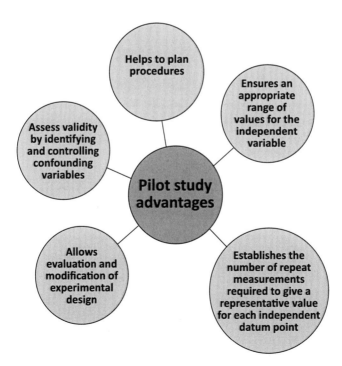

Diagram showing the advantages of conducting a pilot study

2.2.1 Pilot study examples

Example 1

A researcher investigating learning in rats built a maze which included one way doors to prevent the rats from retracing their steps. Before conducting the formal investigation she carried out a pilot study using two rats. Whilst carrying out the pilot study she noticed that the one way doors did not stay open long enough to allow the rats' tails to pass through. The rats learned that going through a door meant their tail would be struck and they began to refuse to enter new sections of the maze. To solve the issue, the researcher placed small wooden blocks at the bottom of the doors to allow the rats' tails to pass through unharmed. As a result of this modification, the rats were not afraid to enter the next section of the maze and the formal investigation was completed without any experimental design flaws.

Laboratory rat
(Source: National Cancer Institute (http://1.usa.gov/1VEI22I))

Example 2

A student was investigating the inhibitory effects of lead on the activity of the enzyme catechol oxidase. The student would be mixing catechol (the substrate) with catechol oxidase and lead ethanoate at varying concentrations, then determining the activity of catechol oxidase. Before beginning the full investigation, the student conducted a pilot study to determine the concentrations of lead ethanoate which should be used in the main study. This allowed the student to ensure she was using an appropriate range of values for the independent variable (lead ethanoate concentration) and avoided the results for the dependent variable going 'off the scale', i.e. 100% inhibition.

Example 3

A group of researchers conducting a study in the Amazon rainforest carried out a small scale investigation in a rainforest bio-dome in the UK to ensure that the electronics they would be using are capable of functioning in a warm humid atmosphere. This was an important step as it would be very costly to send a whole team and equipment to the Amazon and discover that the equipment they required was not functional due to the humidity.

2.3 Experimental design

This section considers independent and dependent variables, confounding variables, controls, in vivo and in vitro studies.

2.3.1 Independent and dependent variables

Any scientific experiment will be subject to a series of variables. Variables are factors which can be changed, controlled or measured within an experiment.

The independent variable is the variable that is changed in a scientific experiment. A dependent variable is the variable being measured in a scientific experiment. Experiments involve the manipulation of the independent variable by the investigator and the measuring of the dependent variable.

Daphnia are a type of plankton whose heart rate can be directly monitored by observation through a microscope. A student carried out an investigation into the effect of water temperature on heart rate in Daphnia. For this experiment, water temperature ($°C$) is the independent variable and heart rate (beats per minute) is the dependent variable.

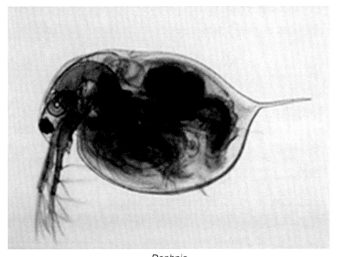

Daphnia
(Daphnia magna (http://bit.ly/1LbHCb6) by Ayacop (http://bit.ly/1QkdylF) is licensed under CC BY 2.5 (http://bit.ly/1em5XTc))

Independent and dependent variables can be continuous (range of values) or discrete (distinct groups). In the Daphnia example, both water temperature and heart rate are continuous.

The difficulty in designing an experiment will vary depending on the type and complexity of the experiment. Simple experiments involve a single independent variable. For example testing the effect of varying concentrations of a new drug on cells in the lab is a simple experiment; there is only one independent variable - the concentration of the drug. The control of laboratory conditions allows simple experiments to be conducted more easily than in the field. However, a drawback of a simple experiment is that its findings may not be applicable to a wider setting.

A multifactorial experiment involves a combination of more than one independent variable or combination of treatments. This is a more complex type of experiment compared to one which has a single independent variable. This seems to be contrary to what has gone before: how can an experiment have more than one independent variable? Consider again the effect of drugs on human physiology. Many drugs alter their effect when combined with other therapies. While examining the effect of one drug on a culture of human cells may provide a single independent variable, the results of this study are less useful if the drug is usually used in combination with one or more other therapies. The following table shows the different treatment groups in a multifactorial experiment investigating the effects of an antidepressant drug (Prozac) in combination with a form of psychotherapy called cognitive behavioural therapy (CBT).

		Drug therapy	
		Placebo	Prozac
Psychotherapy	None	Control	Prozac only
	CBT	CBT only	Combined therapy

Multifactorial experiment

At the end of the experiment the severity of depression can be measured (using the Beck Depression Inventory (BDI)) and the results between the different treatment groups can be analysed. In this case conducting a multifactorial experiment provides more robust findings into the benefits of combining psychotherapy and drug therapy in treating depression rather than looking at each factor individually. In general, as experimental designs increase in complexity, more information can be obtained and analysed as well as allowing the detection of interaction effects.

Observational studies use groups that already exist to carry out an investigation. For example a study investigating the relationship between smoking and lung cancer would recruit an experimental group of people who have been smoking for a varying number of years and a control group of people who have never smoked. This is the only viable method to recruit participants to this investigation as it would be unethical to ask a group of individuals to start smoking cigarettes to allow a comparison to a control group. Recruiting experimental groups in this way also allows researchers to study long-term effects of certain variables (i.e. effects lasting decades) which may not otherwise be possible.

Because observational studies use groups that already exist, the researcher is not able to tightly control the composition of the experimental groups; this means there is the potential for bias in the groups to affect the results of the study and there is no truly independent variable. As a result, these 'observational' studies are good at detecting correlation but, as they do not directly test the model, they are less useful for determining causation.

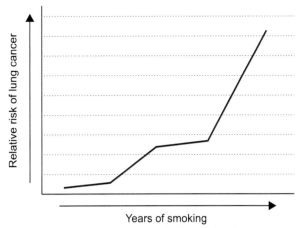
Correlation between smoking and relative risk of lung cancer

Observational studies can be very powerful, but the weakness is that all variables cannot be controlled for. When a correlation is found in observational studies - that is when the assumption of cause and effect must be avoided, and more thorough analysis is required. If A correlates with B, then A may cause B, B may cause A, A and B may be caused by a common variable C, or the correlation may be a statistical fluke and not 'real'. Further studies are then required to confirm the correlation and any specific causal hypothesis. In the smoking example mentioned earlier researchers suggested that there was a correlation between smoking and increased incidence of lung cancer; however, to determine causation further evidence was gathered such as:

- stopping smoking reduces the risk of cancer;
- greater intensity of smoking increases risk;
- tobacco smoke contains substances demonstrated to be carcinogens - so there is biological plausibility.

2.3.2 Confounding variables

Due to the complexities of biological systems, other variables besides the independent variable may affect the dependent variable. These variables are known as confounding variables. A confounding variable is any factor which affects the dependent variable that is not the independent variable. These confounding variables must be held constant if possible, or at least monitored so that their effect on the results can be accounted for in the analysis.

In the experiment investigating the effect of water temperature on heart rate in Daphnia there will be many confounding variables for example age of Daphnia, size of Daphnia, light intensity of surroundings, pH of water, mineral content of water and method of counting heartbeat.

In cases where confounding variables cannot easily be controlled, blocks of experimental and control groups can be distributed in such a way that the influence of any confounding variable is likely to be the same across the experimental and control groups. In randomised block design experimental subjects are assigned into groups or 'blocks' before being assigned to a treatment group. For example the size of the Daphnia has been identified as a confounding variable in the investigation described in section 2.3.1. To carry out a randomized block design, the Daphnia would first be placed into groups according to their size. Then from each group, individuals would be assigned to the different temperature treatment groups. In this way the influence of Daphnia size is likely to be the same across the experimental and control groups.

In another example scientists conducted an experiment to investigate the effects of three different fertilisers on the growth of plants. The plants were grown in a greenhouse. Depending on their position in the greenhouse, the plants would be exposed to varying micro-climates (for example in terms of light intensity and humidity) which cannot be controlled; therefore, a randomised block design was used as shown in the following diagram.

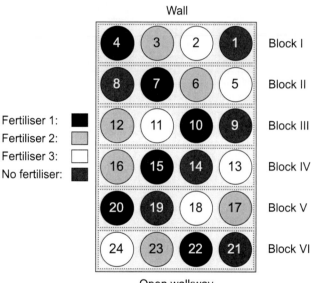

Randomised block design

2.3.3 Controls

Control experiments are an essential aspect of all valid scientific research projects. The results of control groups can be used to determine if an experiment is working properly at a procedural level and are also used for comparison with treatment results. Experiments usually employ one (or both) of the following types of control:

- positive control;
- negative control.

A positive control is a treatment that is included to check that the system can detect a positive result when it occurs. This group is expected to have a positive result and proves to the researcher that the experimental design being used is capable of producing results.

Imagine an experiment investigating the effect of a newly discovered antibiotic on the growth of E. coli bacteria. Petri dishes containing agar and different concentrations of the new antibiotic would be set up alongside a positive control dish containing agar with an antibiotic which is known to inhibit the growth of E. coli. A known concentration and volume of E. coli bacteria would be spread on the agar surface of each petri dish and incubated for 24 hours. If growth is observed on all the petri dishes, except the positive control, this suggests that the new antibiotic is not capable of inhibiting the growth of E. coli. If growth is observed on all the petri dishes, including the positive control, this suggests there is a flaw in the experimental design or procedure. Positive scientific control groups reduce the chances of false negative results.

A negative control group provides results in the absence of a treatment. If a negative control yields an unexpected result this could suggest that there is an unknown confounding variable affecting the experiment and therefore the results would be considered invalid.

In the antibiotic example mentioned earlier, a petri dish containing agar only would be set up to act as a negative control. A known concentration and volume of E. coli bacteria would be spread on the agar surface of the petri dish and incubated for 24 hours. If inhibition of growth is observed on all the petri dishes, except the negative control, this suggests that the new antibiotic is capable of inhibiting the growth of E. coli. If inhibition of growth is observed on all the petri dishes, including the negative control, this suggests there is another variable affecting the inhibition of growth and the results are invalid. Negative scientific control groups reduce the chances of false positive results.

Placebos can be included as a treatment without the presence of the independent variable being investigated. The placebo effect is a measurable change in the dependent variable as a result of a patient's expectations, rather than changes in the independent variable. Placebos are often used in drug trials where one group of patients is given a new drug and the other group is given a placebo. In a double blind drug trial neither the patient nor the researcher know which patients are receiving the drug and which are receiving the placebo.

A double blind experiment was conducted on patients with Alzheimer's disease. The experiment was testing the effects of extract of Ginkgo biloba (EGb) on the patients' cognitive abilities. The patients' cognitive abilities were measured using the Alzheimers Disease Assessment Scale - cognitive subscale (ADAS-Cog); they were then given extract of Ginkgo biloba or a placebo to take for one year after which their cognitive abilities were re-measured. The results are shown in the following table. (Note: an improvement in performance is indicated by a decrease in ADAS-Cog score.)

	Change in ADAS-Cog score				
	-4 or better	-3 to -2	-1 to +1	+2 to +3	+4 or worse
EGb	22	18	12	7	16
Placebo	10	11	19	11	24

For each of the following experiments, read the description of each experiment and the controls it included.

Controlled experiment 1

A researcher carried out a clinical trial to investigate the effectiveness of a new medication which controls asthma symptoms. A group of asthmatic subjects were randomly assigned to one of three groups, one experimental group and two control groups. Subjects were told to use a specific inhaler twice a day over a period of six weeks and take a peak flow reading morning and afternoon each day. Peak flow is a measure of the fastest airflow that can be blown from the lungs.

Decide if the description of each of the following control groups indicates a positive or negative control.

Q1: In one control group, the subjects were given an inhaler containing a medication which has already proved to be effective in reducing asthma symptoms.

a) Positive
b) Negative

....

Q2: In one control group, the subjects were given an inhaler containing a **placebo**.

a) Positive
b) Negative

Controlled experiment 2

A researcher carried out an enzyme assay to determine the quantity of enzyme in an extract. The reaction rate of the enzyme was determined by monitoring the accumulation of the product with time. The quantity of enzyme present was calculated from the observed reaction rate.

Decide if the description of each of the following control groups indicates a positive or negative control.

Q3: A control was set up containing no enzyme.

a) Positive
b) Negative

....

Q4: A control was set up containing a known quantity of the purified enzyme.

a) Positive
b) Negative

2.3.4 In vitro vs. in vivo experiments

Experiments can be carried out in vitro or in vivo. In vitro refers to the technique of performing a given procedure in a controlled environment outside of a living organism e.g. cells growing in culture medium, proteins in solution, purified organelles. An in vivo experiment describes an investigation which is carried out using a whole, living organism, e.g. mouse model organism.

In vitro   *In vivo*

(Cell Culture in a tiny Petri dish (http://bit.ly/1IID3pY) by kaibara87 (http://bit.ly/1WOOk37) is licensed under CC BY 2.0 (http://bit.ly/1rRyEZO),
Laboratory mouse (http://bit.ly/1GYpW85) by Rama (http://bit.ly/20OGm9s) is licensed under CC BY-SA 2.0 FR (http://bit.ly/1LbNXU7))

An in vitro experiment may be elegant and easily controlled but its relevance may be limited in vivo. In vivo experiments allow researchers to investigate the overall effect of an agent on a living organism. For example in vitro testing using cells in culture may be used as an initial safety test of a new therapeutic compound; however, in vivo testing using an animal model must be performed before advancing to human clinical trials as this will allow the investigators to assess the performance of the drug within a biological system.

The following table outlays the advantages and disadvantages of in vitro and in vivo experiments.

In vitro	Invivo
Less expensive	Expensive
Less time-consuming	More time-consuming
Less precise	More precise
Incomplete representation of the in vivo system	Accurately represent the in vivo situation

2.4 Sampling

For many investigations, sampling an entire population (or ecosystem) simply is not feasible. This may be due to lack of money, time constraints, lack of equipment, sheer number of subjects etc. When conducting an investigation, an appropriate sampling strategy must be used. This allows the researcher to select a representative sample of the population and use the results to reach conclusions about the population as a whole.

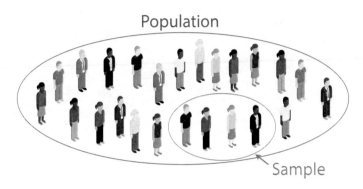

Sampling

A challenge which all researchers face is determining an appropriate sample size for an investigation. A large sample may yield more reliable results; however, correlations can be lost due to 'noise' in the data. In general, the extent of the natural variation within a population determines the appropriate sample size. The more variable a population is, the larger the sample size which is required. A representative sample should share the same mean and the same degree of variation about the mean as the population as a whole.

The degree of variation within a population is often determined by calculating standard deviation (SD). Standard deviation is a measure which allows the amount of variation in a set of data to be quantified. In general a larger sample size will yield a standard deviation which is closer to that of the population as a whole. Correspondingly, a larger sample size should also give a more accurate estimation of the population mean. Calculating the true mean or SD of a whole population can also be challenging as it may not be possible to determine the true value for each individual within the main population. Ultimately all researchers must remember that the data gathered from a sample group only represents a subset of the whole population; therefore, any conclusions drawn from the investigation are subject to a margin of error.

To achieve a representative sample group researchers may employ one of three different sampling strategies:

- random sampling;
- systematic sampling;
- stratified sampling.

Random sampling

In random sampling, members of the population have an equal chance of being selected. This reduces the possibility of bias in the sampling group. When sampling a population, the individuals are numbered and selected at random to form the sample group (as shown in the following diagram). When sampling an ecosystem a grid can be drawn over a map of the area of study and random coordinates are selected to determine the location of the sampling points.

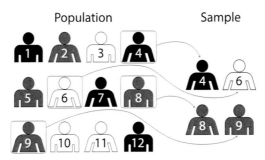

Random sampling

The advantages of random sampling are that it is a straightforward process and avoids bias in the sample group. One disadvantage of random sampling is that it can result in poor representation of the main population/area if certain members/areas are not selected by the random numbers generated. There may also be issues with accessing all the selected members of the population or sites from an area of study.

Systematic sampling

In systematic sampling, members of a population are selected at regular intervals. For example members of a population may be listed and every fourth individual selected to form part of the sample group. In an environmental study, systematic sampling may involve samples being taken in a regular pattern, i.e. every three metres along a transect line.

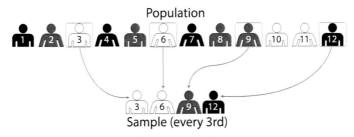

Systematic sampling

The advantage of systematic sampling lies in the fact that it should provide a more representative sample of the population as a whole (compared to random sampling). However, as a result the sample may be biased because all members (or areas) do not have an equal chance of being selected.

Stratified sampling

In stratified sampling, the population is divided into categories that are then sampled proportionally. This means the population is organised into groups or 'strata' according to some characteristic (e.g. age) and the number of individuals sampled from each group is in proportion to the group's size in the main population. In an environmental study this may mean an area is split into separate habitats and proportionally sampled.

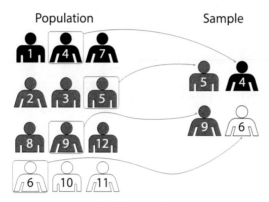

Stratified sampling

The main advantage of stratified sampling is that it should give a sample which is highly representative of the population as a whole. The difficulty with the use of stratified sampling is that the proportions of each group within the main population must be known. In order to identify the relative proportion of each group, the researcher must have access to accurate, up-to-date population data and this is not always available.

2.5 Reliability

When variation arises in experimental results, this may be due to the reliability of measurement methods. To determine the reliability of measuring instruments being used in an investigation, a researcher can take repeated measurements or readings of an individual datum point. The variation observed in the results indicates the precision of the measurement instrument or procedure but not necessarily its accuracy.

For example, in an experiment using a colorimeter to take an absorbance reading of a solution a researcher may take three absorbance readings of a solution and find that the readings are all very similar. This indicates that the results are precise because there is little variation around the mean value. The results, however, could still be inaccurate. For example if the cuvette being used to hold the solution was not clean the absorbance readings would be affected and the results would be inaccurate.

Variation in experimental results may also be due to the inherent variation in the specimens. The natural variation in the biological material being used can be determined by measuring a sample of individuals from the population. The mean of these repeated measurements will give an indication of the true value being measured. Using a larger sample size should help to produce a more representative result by showing the full range of data present within the population. Using the data from a large sample group should then provide the most accurate mean value for the characteristic being studied.

For example in an investigation into the effect of population density on average shell length of limpets, researchers may find large variation in their repeat measurements.

Limpets
(Common limpets (http://bit.ly/1NDmVg4) by Tango22 (http://bit.ly/1QOZwsl) is licensed under CC BY-SA 3.0 (http://bit.ly/1kvyKWi))

This variation may arise due to the reliability of measurement methods, i.e. measuring limpet shell length using a ruler with mm subdivisions may not provide an accurate result. It may also be difficult to accurately measure the shell length due to its uneven shape. Genetic differences between limpets may result in added variation in shell length data, thus increasing variation within the results even further. Altogether the variation due to measurement methods and the inherent variation within the population will reduce the reliability of the results. Increasing the sample size would provide the most representative results and allow a more accurate mean to be calculated.

Reliability of results is something which all research scientists strive to achieve. Results are considered to be reliable when they can be achieved consistently both by the initial researcher and other scientists following the same procedure. The main reason behind the need for reliability is that in order for results to be considered to be significant, they must be repeatable, not just a one off result. By ensuring results are both reliable and valid, findings are likely to be accepted as true by the scientific community.

2.6 Presentation of data

Whether a variable is discrete or continuous, it can give rise to qualitative, quantitative or ranked data. Quantitative can be measured objectively, usually with a numerical value. In other words it refers to information which can be measured with numbers, e.g. volume, mass, time, temperature etc. Qualitative data is subjective and descriptive. In this case the information can be observed but cannot actually be measured, e.g. case studies and interviews.

Some experiments may also give rise to ranked data. Ranked data refers to the data transformation in which numerical values are replaced by their rank when the data are sorted from lowest to highest. One form of ranked data places observations into an order from smallest to largest (or vice versa). For example a group of researchers observing six male colobus monkeys in a zoo were attempting to determine the dominance hierarchy which existed within the group.

Colobus monkey
(Colobus Monkey at the Oregon Zoo (http://bit.ly/1O5nmwE) by Cacophony (http://bit.ly/1MD jgwq) is licensed under CC BY-SA 3.0 (http://bit.ly/1kvyKWi))

By observing the pushing behaviour of the monkeys, the researchers were able to establish a rank order of dominance amongst the six male monkeys, from most dominant to least dominant. The results are shown in the following table.

Monkey name	Dominance rank
Aldo	1
Bandar	2
Clyde	3
Yono	4
Ari	5
Virgil	6

Ranked data

Another form of ranked data involves sorting data into order and replacing each value with a number to represent its position in the sequence. For example, following on from the previous experiment to determine the rank order of dominance in the group of male monkeys, researchers wanted to investigate the relationship between **parasite load** and dominance. They counted the number of *Trichuris* worm eggs per gram of faeces in each of the monkeys to determine parasite load. This quantitative data was then converted into ranked data shown in the following table. Using the information in the table, statistical tests can be performed to determine if the two variables (dominance rank and parasite load) covary.

Monkey name	Dominance rank	Egg per gram of faeces	Eggs per gram (rank)
Aldo	1	5678	1
Bandar	2	4333	2
Clyde	3	2836	3
Yono	4	1642	4
Ari	5	793	6
Virgil	6	843	5

Ranked data

The type of variable being investigated has consequences for any graphical display or statistical tests that may be used. For example experiments involving discrete variables display results using bar graphs whereas experiments involving continuous variables display results using line graphs. When graphing data, correlation exists if there is a relationship between two variables. Correlation is an association and does not imply causation. Causation exists if the changes in the values of the independent variable are known to cause changes to the value of the dependent variable. A positive correlation exists when an increase in one variable is accompanied by an increase in the other variable. A negative correlation exists when an increase in one variable is accompanied by a decrease in the other variable. Strength of correlation is proportional to spread of values from line of best fit. In other words when the values align closely to the line of best fit the correlation is described as being strong, when the values do not align closely to the line of best fit the correlation is described as being weak.

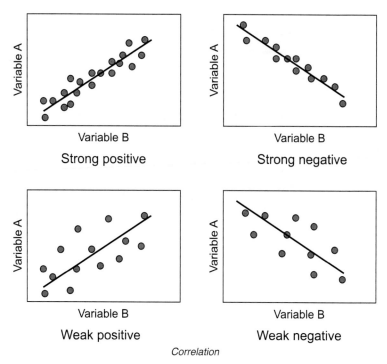

Correlation

Statistical tests can be conducted on data including calculations of mean, median and mode. The mean is the average value. The median is the middle value when the data is placed in sequence. The mode is the most frequent value in the data set.

Box plots can be used to show variation within and between data sets. Box plots should include:

- median;
- lower quartile;
- upper quartile;
- inter-quartile range.

The following box plot was generated from data on the shoe size of a group of pupils.

TOPIC 2. EXPERIMENTATION

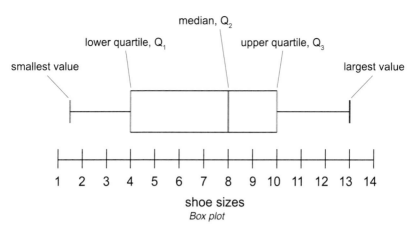

Box plot

The box plot has five important components which help us to analyse the data presented. Firstly the smallest value and largest value indicate the range of data which was gathered, this is one straightforward method for demonstrating the spread of the data.

Another measure of the spread of the data which we can determine from a box plot is the interquartile range. This splits the data into quarters. To find the interquartile range we subtract the lower quartile from the upper quartile. So, reading from the boxplot we get:

$$\begin{aligned}\text{Interquartile range} &= Q3 - Q1 \\ &= 10 - 4 \\ &= 6\end{aligned}$$

Analysis of the interquartile range is the preferred method for determining the spread of the data since it is unaffected by any outliers - data points which sit far away from all the others.

Finally, there is also the median. The median refers to the value that is right in the middle of the data. The median is the central line in the box plot, so we simply read it off to be 8.

It is often useful to compare two box plots of the same information collected about two different groups, e.g. the heights of men compared to the heights of women. When doing this, all you need to do is state which one has a greater spread than the other (by looking at the IQR and/or the range) and which one has a higher average (by looking at the median).

Example : Box plot example

Construct a box plot for the following data set.

$$3, 5, 8, 8, 9, 11, 12, 12, 13, 13, 16$$

Finding the largest and smallest terms is easy: they are 3 and 16 respectively. Finding the lower quartile, median, and upper quartile is bit more effort. To find the median, you may recall that if we add 1 to the total number of objects and then divide by 2, that tells us the median. In general, we say the median is the $\frac{n+1}{2}$th term, where n is the total number of objects we're working with.

We divide by 2 because the median is half way through the data. Considering that the lower quartile is a quarter of the way through and the upper quartile is 3 quarters of the way through, we get:

- the lower quartile is the $\frac{n+1}{4}$ th term;
- the upper quartile is the $\frac{3(n+1)}{4}$ th term.

This set of data contains 11 numbers, so we get the following:

- the median is the $\frac{11+1}{2}$ = 6th term which is 11;
- the lower quartile is the $\frac{11+1}{4}$ = 3rd term so Q1 = 8;
- the upper quartile is the $\frac{3(11+1)}{4}$ = 9th term, so Q3 = 13;

Now we have all the information we need to draw a box plot.

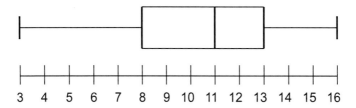

2.7 Extended response question

The activity which follows presents an extended response question similar to the style that you will encounter in the examination.

You should have a good understanding of pilot studies before attempting the question.

You should give your completed answer to your teacher or tutor for marking, or try to mark it yourself using the suggested marking scheme.

Extended response question: Pilot studies

Discuss the advantages of building a pilot study into the development of a biological investigation. *(3 marks)*

2.8 Learning points

Summary

- Validity: variables controlled so that any measured effect is likely to be due to the independent variable.
- Reliability: consistent values in repeats and independent replicates.
- Accuracy: data, or means of data sets, are close to the true value.
- Precision: measured values are close to each other.
- Integral to the development of an investigation, a pilot study is used to help plan procedures, assess validity and check techniques; this allows evaluation and modification of experimental design.
- The use of a pilot study can ensure an appropriate range of values for the independent variable; in addition, it allows the investigator to establish the number of repeat measurements required to give a representative value for each independent datum point.
- An independent variable is the variable that is changed in a scientific experiment.
- A dependent variable is the variable being measured in a scientific experiment.
- Independent and dependent variables can be continuous (having a range of values) or discrete (falling into distinct groups).
- Experiments involve the manipulation of the independent variable by the investigator.
- The experimental treatment group is compared to a control group.
- Simple experiments have one independent variable, they are straightforward to perform (due to the ease of controlling laboratory conditions) but their results may not be applicable to an in vivo situation.
- Multifactorial experiments have more than one independent variable or combination of treatments, they are more complex to perform and analyse but their results are more applicable to an in vivo situation.
- Investigators may use groups that already exist, so there is no truly independent variable, these are known as observational studies.
- Observational studies are good at detecting correlation, but since they do not directly test a hypothesis, they are less useful for determining causation.
- In observational studies the independent variable is not directly controlled by the investigator, for ethical or logistical reasons.
- Due to the complexities of biological systems, other variables besides the independent variable may affect the dependent variable; these are known as confounding variables.
- These confounding variables must be held constant if possible, or at least monitored so that their effect on the results can be accounted for in the analysis.

> Summary continued

- In cases where confounding variables cannot easily be controlled, a randomised block design could be used.
- Randomised blocks of treatment and control groups can be distributed in such a way that the influence of any confounding variable is likely to be the same across the treatment and control groups.
- Control results are used for comparison with the results of treatment groups.
- The negative control provides results in the absence of a treatment.
- A positive control is a treatment that is included to check that the system can detect a positive result when it occurs.
- Placebos can be included as a treatment without the presence of the independent variable being investigated.
- Placebo effect is a measurable change in the dependent variable as a result of a patient's expectations, rather than changes in the independent variable.
- In vitro refers to the technique of performing a given procedure in a controlled environment outside of a living organism.
- Examples of in vitro experiment include: cells growing in culture medium, proteins in solution, purified organelles.
- In vivo refers to experimentation using a whole, living organism.
- In vitro experiments may be more straightforward to perform than in vivo experiments; however, their results may be less applicable to a wider setting.
- Where it is impractical to measure every individual, a representative sample of the population is selected.
- The extent of the natural variation within a population determines the appropriate sample size; more variable populations require a larger sample size.
- A representative sample should share the same mean and the same degree of variation about the mean as the population as a whole.
- In random sampling, members of the population have an equal chance of being selected.
- In systematic sampling, members of a population are selected at regular intervals.
- In stratified sampling, the population is divided into categories that are then sampled proportionally.
- Variation in experimental results may be due to the reliability of measurement methods and/or inherent variation in the specimens.
- The reliability of measuring instruments or procedures can be determined by repeated measurements or readings of an individual datum point; the variation observed indicates the precision of the measurement instrument or procedure but not necessarily its accuracy.

Summary continued

- The natural variation in the biological material being used can be determined by measuring a sample of individuals from the population; the mean of these repeated measurements will give an indication of the true value being measured.
- The range of values is a measure of the extent of variation in the results; if there is a narrow range then the variation is low.
- Independent replication should be carried out to produce independent data sets.
- These independent data sets should be compared to determine the reliability of the results.
- Overall results can only be considered reliable if they can be achieved consistently.
- Discrete and continuous variables give rise to qualitative, quantitative, or ranked data.
- Qualitative data is subjective and descriptive.
- Quantitative data can be measured objectively, usually with a numerical value.
- Ranked data refers to the data transformation in which numerical values are replaced by their rank when the data are sorted from lowest to highest.
- The type of variable being investigated has consequences for any graphical display or statistical tests that may be used.
- The mean is the average value in a data set.
- The median is the middle value when the data is placed in sequence.
- The mode is the most frequent value in the data set.
- Use box plots to show variation within and between data sets including identification and calculation of the median, lower quartile, upper quartile and inter-quartile range.
- Correlation exists if there is a relationship between two variables; correlation is an association and does not imply causation.
- Causation exists if the changes in the values of the independent variable are known to cause changes to the value of the dependent variable.
- A positive correlation exists when an increase in one variable is accompanied by an increase in the other variable.
- A negative correlation exists when an increase in one variable is accompanied by a decrease in the other variable.
- Strength of correlation is proportional to spread of values from line of best fit i.e. how closely the points lie to the line of best fit.

2.9 End of topic test

End of Topic 2 test   Go online

Q5: Match each of the following terms to its correct definition in the following table: *(4 marks)*

- Accuracy;
- Precision;
- Reliability;
- Validity.

Term	Definition
	The closeness of repeated measurements to one another.
	A measure of how close the data is to the actual true value.
	Refers to the control of confounding variables.
	Refers to whether a procedure yields consistent results each time it is repeated.

....

Q6: An experiment was conducted to investigate the effect of increasing work rate on oxygen uptake in an adult male. The subject was asked to run on a treadmill at a constant speed. The work rate (measured in watts) was increased periodically by increasing the incline of the treadmill. The results of the experiment are shown in the following table.

Work rate (watts)	Oxygen uptake (cm^3 kg^{-1} min^{-1})
50	10
75	15
100	24
125	31
150	36
175	42
200	45

Which is the dependent variable in the experiment? *(1 mark)*

a) Work rate
b) Oxygen uptake

Q7: An experiment was conducted into habitat selection by the sea slug *Onchidoris bilamellata*. The study investigated the effect of both light intensity and texture of surface on habitat selection. The results showed that the sea slugs preferred rough surfaces rather than smooth, however, their surface preference was always overridden by their preference for darkness over light.

What name is given to an experiment, such as this, which has more than one independent variable? *(1 mark)*

....

Q8: Male magnificent frigatebirds (*Fregata magnificens*) have a large red pouch on their throats which they use to make a thrumming sound to attract a mate.

Researchers conducted a study to determine if pouch size was related to the frequency of sound produced. Results from the study are shown in the following scatterplot.

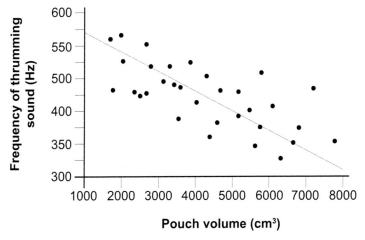

Scatterplot showing the results from the magnificent frigatebird study

Explain why the results are thought to have low significance. *(1 mark)*

....

Q9: A student carried out an investigation into the effect of caffeine on reaction time. She obtained a list of every pupil in her year group and selected every fifth person to take part in the experiment. What type of sampling does this demonstrate? *(1 mark)*

An experiment was conducted to investigate the effect of different metal ions on lipase enzyme activity. The effect of magnesium, cobalt, calcium, potassium, copper and zinc ions was investigated. The enzyme was extracted and purified from a bacterium called *P. aeruginosa*. The enzyme solution was mixed with 1mM of the various ions (as chloride salts) or distilled water (to act as a control) and incubated in a water bath at 30°C for one hour. After the allotted time, the enzyme activity was measured and calculated relative to control activity.

The results of the experiment are shown in the following table.

Metal ion	Enzyme activity (% of control)
Control	100
Magnesium	136
Cobalt	89
Calcium	81
Potassium	79
Copper	73
Zinc	68

Q10: Complete the following sentences, choosing the correct option from the two available options in brackets. *(5 marks)*

This experiment was conducted (*in vitro/in vivo*). The (*dependent/independent*) variable (metal ions) is a (*discrete/continuous*) variable, while the (*dependent/independent*) variable (enzyme activity) is a (*discrete/continuous*) variable.

Q11: What type of control is described in the preceding information? *(1 mark)*

....

Q12: Confounding variables, such as temperature, were tightly controlled throughout the experiment. Explain what is meant by the term confounding variable. *(1 mark)*

....

Q13: Describe the results of the experiment. *(1 mark)*

....

Q14: What evidence is there to suggest that the experiment was only performed once for each type of ion? *(1 mark)*

Unit 3 Topic 3

Critical evaluation of biological research

Contents

3.1 Evaluating background information 287
3.2 Evaluating experimental design 288
3.3 Evaluating data analysis 289
3.4 Evaluating conclusions 293
3.5 Learning points 296
3.6 End of topic test 298

Prerequisites

You should already know that:

- an evaluation is an important part of a scientific report that outlines sources of error, validity of procedures and any possible improvements;
- reliable and relevant sources are used to support background information;
- results and data can be presented in a variety of ways, most commonly as graphs and tables;
- simple calculations such as averages can be used when interpreting data;
- conclusions should refer back to the aim of the investigation.

Learning objective

By the end of this topic, you should be able to:

- state that all scientific reports require:
  - an informative title;
  - abstract or summary - aims and findings;
  - introduction which contains enough detail and context to allow the reader to understand the rest of the report;
- explain that references and citations should be used to support statements made in the report, and describe the standard format used for citing and referencing;
- explain that a method should be present and detailed enough to allow another scientist to repeat the experiment;
- explain that it is important for the procedure to test the aim or hypothesis;
- state that key variables need to be controlled throughout the experiment in order to allow valid conclusions to be made;
- explain that the sample should be random, where possible and appropriate, and of suitable size to allow an unbiased conclusion to be drawn that the independent variable had an effect;
- describe means by which data may be presented, including graphs and tables;
- explain that calculating the mean, median, mode, standard deviation or range of data, where appropriate, can be useful in the interpretation of data;
- state that care must be taken when interpreting results from one-off data or any data that fails to follow the trend;
- describe what error bars and confidence intervals are;
- state that error bars drawn on graphs and the use of confidence intervals can be used to show whether the effect caused by the changing the independent variable may be considered as significant or not;
- state that a correlation does not always imply causation;
- state that the conclusion(s) should refer back to the original aim and hypothesis;
- explain that the suitability, accuracy, validity and reliability of the procedures used in the experiment must be evaluated;
- state that only when the procedure has been deemed valid and results considered significant, can a valid conclusion be drawn;
- explain that conclusions must also be backed up with findings from other reliable investigations that have also been peer reviewed and determined to be valid.

3.1 Evaluating background information

Scientific reports follow this standard format:

1. Title - this should be descriptive and helpful.
2. **Summary** or abstract - short paragraph outlining the aims and findings.
3. Introduction - a concise account providing clear, relevant and unambiguous background information of a suitable depth and detail, whilst justifying the purpose of the study.
4. Methods - clear and detailed to allow easy repetition by a peer.
5. Results - these should be laid out in an appropriate format, e.g. tables with graphs.
6. Discussion - both procedures and results should be evaluated here, allowing valid conclusions to be formed.

The introduction should:

- provide the reader with sufficient information to understand specific aspects of the methods, results and ultimately the discussion;
- provide convincing justification for the study;
- place the study in the context of what is already known and understood about the topic;
- review any key points which both support and contradict the information provided;
- contain several sources to support statements, and citations and references should be in a standard form;
- give details about the ethical considerations behind the decisions made when selecting the particular study methods and organisms;
- contain clear aims and hypotheses.

Evaluating background information　　　　　　　　　　　　　　Go online

Q1: Give two requirements of the summary.

Q2: In the introduction, what should the writer include to support statements?

Q3: A scientist is using *Drosophila melanogaster* in an experiment. What comments regarding this organism should they give in their introduction?

3.2 Evaluating experimental design

In order to make a conclusion, the experiment procedure must be evaluated and deemed to be valid and reliable. The following should be considered:

- **procedures** - these should test the aim or hypothesis;
- **controls** - should be present where appropriate in order to determine that any effects are the result of the treatment or show the effect in the absence of a treatment;
- **controlled variables** - the validity of an experiment may be compromised where factors other than the independent variable may have influenced the dependent variable. A good example of this is ensuring pH remains constant using buffers and temperature remains constant using a water bath;
- **sample size** - must be large enough to state without bias that any effect on the dependent variable was due to changing the independent variable. Of course we already know that in some experiments there may be no effect, and that a negative result can be just as valid. When taking a sample this must be done randomly to ensure that the sample is representative of the entire population, thus preventing **selection bias**;
- **repetition** - experiments must then be repeated in full at a different time using new ingredients to determine reliability.

Evaluating experimental design　　　　　　　　　　　　　　　　　　　　Go online

Q4: Put the following terms into the correct row in the table below:

- Controlled variables;
- Controls;
- Repetition;
- Sample size;
- Selection bias.

Term	Definition
	Experiment set up to show that any effects are the result of the treatment
	Variables that may influence the dependent variable should remain constant throughout the experiment
	Large enough to allow a valid conclusion that the independent variable did have an effect on the dependent variable
	Carrying out the experiment again at a different time with new ingredients to determine reliability
	Where individuals, groups or data are not selected randomly, therefore failing to provide a representative sample

3.3 Evaluating data analysis

In results, data should be presented in a clear, concise and logical manner that permits analysis. This will often be in the form of a graph supported by a table. Raw data should be present in an appendix.

Quantitative or qualitative?

Most experiments involve measuring variables and will therefore be based on **quantitative data** (numbers). Some experiments, however, involve making judgements based on observations. The data in this case will be descriptive and is termed **qualitative**.

Statistics

Data can be further analysed through the appropriate use of simple statistical procedures including:

- **graphs** - should be appropriate type with suitable scales, labels and be plotted accurately. These can be computer generated, however, sometimes the computer program does not produce these to the standard required and drawing by hand is often more suitable. Where possible, **error bars** (95% **confidence intervals**) should be included.
- **mean** - this is the arithmetical average. Add all the values and divide by the number there are.
- **median** - this is the middle point from the range of values. In some cases it may be the same or similar to the mean. If the data is skewed the median may be quite different from the mean.
- **mode** - this is the most common value in a set of data.
- **range** - this is the set of values that the data falls into, thus the smallest value to the biggest value and everything in between, e.g. in an experiment where the wing spans of garden birds were measured the range is the smallest wing span to the largest wing span and all wing spans between.
- **standard deviation (SD)** - shows how much variation or spread from the average/mean exists. The lower the standard deviation the closer the data points are to the mean. Where the standard deviation is higher, this means that the data is spread over a much larger range of values.

Working out the standard deviation is particularly useful when comparing two or more sets of data with the same mean but different ranges. For example, the mean of the following two sets of data is the same: 15, 15, 15, 14, 16 and 2, 7, 14, 22, 30. However, the second is clearly more spread out, thus has a greater range.

Where data is inconsistent and irregular, or where there is a large range and high standard deviation, the validity of the results may be called into question. There are a range of statistical tests that can be used to assess how likely results are to have occurred by chance or if they are in fact significantly different.

Confidence intervals or error bars are indicated in tables or on graphs, showing the spread of data around a mean. Usually, a valid conclusion can be drawn where the average value in the control experiments is significantly different from the average when a treatment has been applied. If the error bars in the treatment experiment overlap with the error bars in the control experiment, it is not usually possible to state with confidence that the treatment has had an effect.

The following data is completely fictional, and has been produced to allow candidates to develop skills in analysing data. The following table shows results for an experiment into the effect of different fertilisers on tomato plant yield.

Fertiliser	Crop yield (kg per hectare)	Standard error with 95% confidence intervals
Supermarket own brand	120	±3.9
Sheep manure	300	±9.4
Chicken manure	180	±4.5
Seaweed extract	340	±10
Tremendous tomatoes	400	±12.7

The following graph shows how these results appear in a graph including error bars.

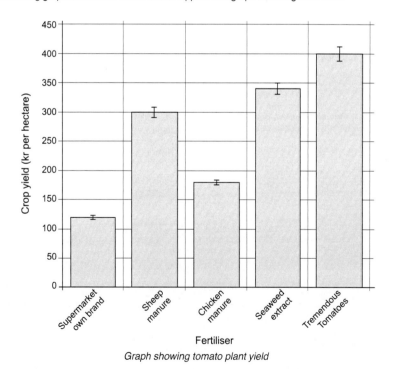

Graph showing tomato plant yield

TOPIC 3. CRITICAL EVALUATION OF BIOLOGICAL RESEARCH

As can be seen from the graph, a valid conclusion can be drawn that tomato plants treated with tremendous tomatoes fertiliser had the greatest yield since error bars do not overlap, and the error bars are not too big.

T-test

The t-test is another statistical test that is useful in determining whether two sets of data are significantly different enough to make a valid conclusion that the treatment has had an effect. The test produces a p-value. Generally if this value is less than 0.05 the null hypothesis can be rejected, thus the treatment has had an effect.

Evaluating data analysis Go online

Q5: Put the following statistical terms into the correct row in the table below:

- Error bar;
- Mean;
- Mode;
- Qualitative;
- Quantitative;
- Range;
- Standard deviation.

Statistical term	Definition
	Data that is measured
	Data that is observed
	The spread or variation of data from the mean
	Average
	Most common value in a set of data
	Full set of data values
	Line drawn on a graph to show the spread of data around the mean

Dog foods experiment

Go online

The following data is completely fictional, and has been produced to allow candidates to develop skills in analysing data. The following table shows the results obtained from an experiment investigating the effect of different dog foods on puppy growth.

Dog food	Increase in mass (kg)	Standard error with 95% confidence intervals
Supermarket own brand	2.2	±0.3
Woof Woof	3.6	±0.2
Bark and be Happy	1.8	±0.1
Wags a lot	4.2	±0.2
Puppy Power	5.6	±0.4

Q6: Present the data in the table as a graph with error bars.

....

Q7: Draw a valid conclusion from the data in table or from your graph.

....

Q8: What about the data allows you able to make this valid conclusion?

Nine husky puppies were in each group. The following is the raw data for Puppy Power and Bark and be Happy.

		Increase in mass for each husky(kg)								
	1	2	3	4	5	6	7	8	9	10
Puppy Power	5.2	6.3	6.4	5.5	5.5	5.4	5.2	5.5	5.6	5.6
Bark and be Happy	1.1	3.4	0.9	1.8	1.2	4.0	0.6	1.1	2.1	2.0

Q9: For which dog food out of Puppy Power and Bark and be Happy, in the table above, is the standard deviation the greatest?

....

Q10: The set of results for which of these two dog foods would provide a conclusion where validity could be questioned and why?

3.4 Evaluating conclusions

In evaluating conclusions, reference should be made to:

- the aim of the study;
- the results obtained;
- the validity and reliability of the experimental design.

Any conclusion should refer back to a hypothesis, stating whether it should be rejected or not. Consideration should be given as to whether significant results can be attributed to **correlation** or **causation**.

Meaningful scientific discussion would include consideration of findings in the context of existing knowledge and the results of other investigations, thus backed by scientific reading. Scientific writing should reveal an awareness of the contribution of scientific research to increasing scientific knowledge and to the social, economic and industrial life of the community. It is here that any significance of the findings are discussed. Do your findings have a potential impact on society, living things or the environment?

Critical analysis of results is an essential part of any conclusion. This is where you take your statistical evidence to state that the conclusion is valid because SD's are small, p-values are small or there is little variation between repeat experiments. You should discuss that there is a big difference between the control group and the experiment group (if there is), making reference to the error bars or confidence intervals. Generally, if the error bars for the control and treatment groups show no overlap, the data may be considered different, thus supporting valid conclusions.

Correlation or causation?

Correlation is where two variables seem to vary together, thus seem to be connected. For example, there is likely to be a correlation between how many sweets a child eats and their incidence of tooth decay.

A **positive correlation** is when an increase in the independent variable seems connected to the increase in the dependent variable. An example of a positive correlation would be that as the external temperature increases, the volume of sweat produced by an individual also increases.

A **negative correlation** is when the dependent variable decreases as the independent variable is increased. An example of a negative correlation would be that as a dairy cow gets older, the volume of milk it produces falls.

A correlation, however, does not always prove that changing the independent variable has directly caused the change in the dependent variable. In short, correlation does not always indicate causation.

> **Example**
>
> A correlation has been found that individuals in some families are more likely to be obese than individuals from other families. This has resulted in some theories surrounding presence of an 'obese gene', thus implying that a gene is causing obesity. Care must be taken when making such a statement.
>
> Obesity is also linked to diet and lifestyle. It is probable that individuals in the same family eat similar foods and follow similar lifestyles. If parents eat a balanced diet and lead active lifestyles, so too do their children. Similarly, where parents eat diets high in saturated fat and sugars, and follow a less active lifestyle, their children often mirror this.
>
> It is therefore, impossible without further study to determine that obesity is caused by a gene, despite there being a clear link between family and obesity.

Positive and negative correlations

Go online

Q11: As chickens get older, the number of eggs laid falls.

a) Positive correlation
b) Negative correlation

....

Q12: As the mass of nitrate increases, the greater the dry mass of algae.

a) Positive correlation
b) Negative correlation

....

Q13: As the level of dissolved oxygen in the river increases so too does the population density of mayfly nymphs.

a) Positive correlation
b) Negative correlation

....

Q14: The longer a person participates in ballet, the larger the surface area of their cerebellum becomes.

a) Positive correlation
b) Negative correlation

....

Q15: The more an athlete trains, the lower their resting pulse rate becomes.

a) Positive correlation
b) Negative correlation

....

Q16: The vitamin C content of broccoli falls as the length of boiling time increases.

a) Positive correlation
b) Negative correlation

The following graph displays results of an experiment to investigate how milk production in cows changes as cows age.

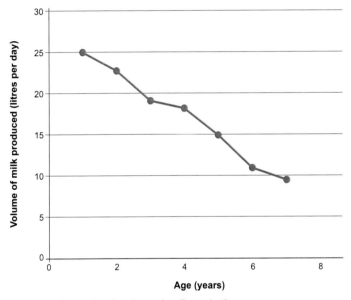

Graph showing change in milk production as cows age

Q17: Is there a correlation between the two variables? If so, is the correlation positive or negative? Give a reason for your answer.

3.5 Learning points

Summary

- All scientific reports should start with an informative title.
- An abstract or summary should follow the title. This states the aims and findings.
- An introduction should follow, providing some background information, enough details to allow the reader to understand the report and an explanation of the context within which the study is set.
- The information in the introduction should be supported by several relevant and reliable sources, correctly cited and referenced.
- The introduction should be followed by the methods section, which should provide clear details which will allow a fellow scientist to repeat the investigation.
- The method should test the aims and hypotheses which should be clearly stated.
- The method should also clearly describe how the independent variable will be changed, how the dependent variable will be measured and how the controlled variables will remain unchanged to maintain validity.
- Within the method, the sample size should be stated, and the manner in which the sample is selected be described.
- A large and unbiased sample group is essential to ensure representative sampling and allow the investigator to show that altering the independent variable has in fact had an effect on the dependent variable.
- A results section will follow the method. This is where data is presented in an appropriate and concise manner.
- Data is usually presented as tables and graphs, with statistical analysis used as appropriate.
- A mean can be calculated from raw data.
- Other statistics such as median, mode and range may also be used.
- Standard deviation can be used to assess the spread of data.
- Scientists should be wary of one-off results or anomalous data.
- Statistical tests such as t-test may be used to determine whether a result is significant or merely due to chance.
- Confidence intervals or error bars can be included in tables or on graphs and show how the data varies from the mean.
- Where the treatment results differ significantly from the control group so that there is no overlap between confidence intervals, the two sets of data can be concluded to be different.
- Conclusions should refer back to the original aim and hypothesis.

TOPIC 3. CRITICAL EVALUATION OF BIOLOGICAL RESEARCH

Summary continued

- For conclusions to be valid, the method must be evaluated in terms validity and reliability. This may include discussing controlled variables, sample size, repetition and accuracy of measurements.

- Conclusions may also be explained or supported through referencing to other studies, again using correct citing and referencing styles.

- Within the conclusion, the economic, social or environmental impact of the findings should be discussed.

- Care should be taken when making conclusions as a correlation between two variables does not always mean that altering one variable has caused the effect on the other.

3.6 End of topic test

End of Topic 3 test — Go online

Q18: A Biology pupil writes the following summary: "The results showed that the seaweed Channel Wrack is found mostly at the top of the shore whereas Dabberlocks is found in the lower shore."
What has the pupil missed out in their summary? *(1 mark)*

....

Q19: Which of the following statements about the scientific method are true? *(3 marks)*

a) It only needs to be a brief outline of the experiment.
b) It should provide sufficient detail to allow repetition by other scientists.
c) It should test the aims and hypotheses.
d) It gives details of variables that should be controlled throughout.

....

Q20: Give four statistical procedures that could be used to analyse data. *(4 marks)*

....

Q21: Complete the sentence below, choosing the correct option from the two available options in brackets. *(3 marks)*

Sample size should be *(large/small)* and selected *(randomly/specifically)* to *(avoid/ensure)* selection bias.

....

Q22: Select the positive correlations from the list below. *(2 marks)*

a) As height increases, the number of spines on holly leaves decreases.
b) As temperature increases, mass of seaweed decreases.
c) As tail length increases, mating success in peacocks increases.
d) As length of time as a gymnast increases, surface area of the cerebellum increases.

....

Q23: In order to reach a valid conclusion, procedures must be .... *(1 mark)*

Unit 3 Topic 4

Investigative biology test

Investigative biology test

Go online

The following experiment and data is fictional. The cited sources are also fictional, added in merely to allow the outcomes to be assessed.

Summary

The main findings of this investigation were that regular medium intensity of exercise for five hours a week significantly lowered resting pulse rate compared to those who did not exercise regularly. More intense or longer duration of exercise showed no further benefit to lowering resting pulse rate.

Background information

The heart is a muscular pump responsible for pumping blood around the body and to the lungs, each time it beats. A healthy resting pulse rate is said to be between 60-80bpm. It has been suggested that a lower resting pulse rate indicates that the heart muscle is stronger and capable of pumping a greater volume of blood with each beat. The heart is therefore under less strain. Pulse can be felt in arteries close to the surface of the skin each time blood is forced from the heart. It has long been reported that participating in regular activity can be a major factor in lowering resting pulse rate (Wilson, 2014), however, the intensity and duration of the exercise required is still unclear in many reports.

This experiment compared and analysed resting pulse rate data from a control group of 30 individuals who did not take regular exercise, with two treatment groups, each of 30 individuals. Treatment group one regularly participated in five hours of moderate intensity exercise per week. Treatment group two regularly participated in eight hours of high intensity exercise per week. In a second experiment, a different 20 individuals who had never participated in regular exercise agreed to exercise for three hours a week for 10 weeks. Their resting pulse rate was measured before and after.

TOPIC 4. INVESTIGATIVE BIOLOGY TEST

Q1: The aim has been left out of the summary. Which of the following would be a suitable aim for this experiment? *(1 mark)*

a) To find out the effect of regular exercise on fitness.
b) To find out the effect of exercise intensity and duration on resting pulse rate.
c) To find out the effect of exercise intensity and duration on mass of heart muscle.

....

Q2: Which of the following provides a suitable null hypothesis for this experiment? *(1 mark)*

a) The greater the level of activity the higher the resting pulse rate.
b) Those who participate in regular activity will have healthier hearts.
c) Exercise intensity and duration will have no effect on resting pulse rate.

....

Q3: This experiment involves working with human subjects over the age of 16.

Which of the following statements are true regarding the need for informed consent? *(3 marks)*

a) Ensures participants know they can withdraw at any time.
b) Ensures participants know that once they sign up they have to complete the study.
c) Allows participants to assess any harm or risk that might be associated with the experiment.
d) Allows participants to assess any benefits to society that may arise from the results of the investigation.
e) Only those over the age of 16 are eligible for informed consent.

....

Q4: In the background information, how did the writer support the idea that regular exercise lowers resting pulse rate? *(1 mark)*

Method

Subjects

The subjects included in the experiment were volunteers aged between 40 and 55 years. All participants received a health check before taking part in the experiment and came from a similar socio-economic background. Each experimental group comprised an even balance of 50% males and 50% females with an average age of 47±1. Age range did not differ between the control and treatment groups.

Experiment 1

In experiment 1, 30 individuals were included in a control group which did not take part in regular exercise. For treatment group 1, 30 individuals were asked to follow an exercise regime which involved five hours of moderate intensity exercise per week. Moderate intensity exercise included fast-paced walking, water aerobics, hiking and gentle bike riding. For treatment group 2, 30 individuals were asked to follow an exercise regime which involved eight hours of high intensity exercise per week. High intensity exercise included jogging/running, fast-paced swimming, aerobics and singles tennis. After 12 weeks, the resting pulse rate of each participant was taken using a digital pulse rate monitor.

Experiment 2

In experiment 2, 20 individuals who had never participated in regular exercise were asked to follow an exercise regime which involved three hours of moderate intensity exercise per week. Moderate intensity exercise included fast-paced walking, water aerobics, hiking and gentle bike riding. The resting pulse rate of each participant was taken using a digital pulse rate monitor before beginning the exercise regime and again 10 weeks later.

Q5: Which of the following best describes the independent variable in experiment 1? *(1 mark)*

a) Age and gender of the participants
b) Whether participants participated in regular exercise or not
c) Pulse rate of participants

....

Q6: Describe the ethical considerations carried out in this investigation. *(1 mark)*

....

Q7: The researcher identified age as a confounding variable. How did they try to ensure the influence of this confounding variable was the same across the experimental and control groups? *(1 mark)*

....

Q8: Experiment 1 investigated both duration and intensity of exercise. What term describes an experiment like this which involves a combination of treatments? *(1 mark)*

....

Q9: In experiment 1, the individuals in the control group did not take regular exercise. Explain how this group acts as a negative control. *(1 mark)*

....

Q10: In both experiments, the researcher used a digital pulse rate monitor to measure the pulse rate of each participant. How could the reliability of the digital pulse rate monitor be determined? *(1 mark)*

Results for experiment 1

The results in the table are mean values of the resting pulse rates of the 30 individuals in each of the three groups, with 95% confidence intervals included.

Group	Mean resting pulse rate (bpm)
Control	98±8
Treatment 1 - moderate intensity exercise for five hours per week	61±3
Treatment 2 - high intensity exercise for eight hours per week	55±14

Q11: What sort of data is present in the table above? *(1 mark)*

....

Q12: Following statistical analysis of the data in the above table, the following statements were made:

- Individuals who participated in moderate intensity exercise for five hours a week had significantly lower resting pulse rates compared to the control group;
- Individuals who participated in high intensity exercise for eight hours a week did not have significantly lower resting pulse rates compared to treatment group 1.

What evidence from the table supports these statements? *(1 mark)*

....

Q13: What type of relationship exists between participation in regular exercise and resting pulse rate? *(1 mark)*

Results for experiment 2

The results in the following table are mean values of the resting pulse rates of the 20 individuals before and after their exercise programme.

Time	Mean resting pulse rate (bpm)
Before	101±6
After	74±4

Q14: Calculate the percentage decrease (to one decimal place) in resting pulse after participating in the exercise programme. *(1 mark)*

....

Q15: In the discussion the investigator writes: *"All individuals in the study were volunteers from a small church group aged 40-55 years old."*

What might a peer be worried about when reading this statement during peer review? *(1 mark)*

TOPIC 4. INVESTIGATIVE BIOLOGY TEST

Citric acid is used as a flavouring agent and is produced commercially using the mold *Aspergillus niger (A. niger)*. Previous experiments have confirmed that *A. niger* is capable of utilising sucrose to produce citric acid, however, little research has been conducted into the utilisation of other sugars. An experiment was therefore conducted to investigate the effect of different types of sugar (glucose, fructose, lactose or galactose) on citric acid production.

Cultures of *A. niger* were grown in a fermenter with growth medium containing either glucose, fructose, lactose or galactose. After 14 days, the citric acid and sugar content of each sample was determined using liquid chromatography. The results are shown in the following table.

Sugar	Sugar utilised (g/l) [a]	Citric acid (g/l)	Citric acid yield (%) [b]
Glucose	88	34	38
Fructose	75	22	29
Lactose	66	7	11
Galactose	70	0	0

[a] Initial sugar concentration 140 g/l
[b] Based on sugar utilised

Q16: Identify the independent variable in this experiment. *(1 mark)*

....

Q17: Use the information above to design a positive control for this experiment. *(1 mark)*

....

Q18: The researchers monitored but did not control the pH in the fermentation vessel because previous experiments have shown that pH is not a confounding variable. Explain what is meant by the term confounding variable. *(1 mark)*

....

Q19: In this investigation, what would be a suitable null hypothesis? *(1 mark)*

....

Q20: Describe the results of the experiment. *(1 mark)*

....

Q21: Explain why the results of this investigation may not be reliable. *(1 mark)*

An experiment was conducted to investigate the effect of caffeine on reaction time. Prior to embarking on the main study, a small scale investigation was conducted and found that the two treatment groups should be administered 500 mg or 250 mg of caffeine.

For the main study, thirty female students aged between 18 and 22 were systematically sampled from a university population (the process was repeated to select thirty male participants). Ten women and ten men were randomly assigned to one of three groups. The experiment was conducted in a double-blind manner. Participants from treatment group 1 were administered 500 mg caffeine, treatment group 2 were administered 250 mg of caffeine and group 3 (the control group) were given a placebo. Reaction time was measured using an electronic test before and after consumption of one of the three solutions. The second reaction time test was conducted one hour after consumption of one of the solutions to allow absorption of the caffeine.

Q22: What term describes an experiment, like the one described in the information above, which allows refinement of experimental design before embarking on the main investigation? *(1 mark)*

....

Q23: Describe the process of systematic sampling. *(1 mark)*

....

Q24: This experiment was conducted in a double-blind manner, meaning both the researchers and the subjects are unaware of which treatment is being administered to each group. Suggest how this feature of the study improves its validity. *(1 mark)*

....

Q25: Identify one confounding variable, not already mentioned, which could affect this experiment. *(1 mark)*

....

Q26: Draw two conclusions from the results of the caffeine reaction time study shown in the following graph. *(2 marks)*

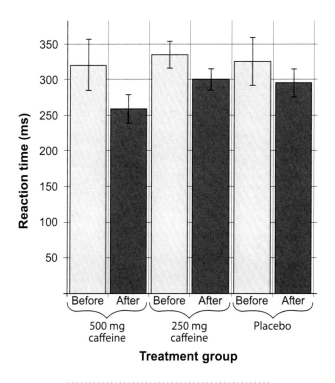

....

Q27: Why would increasing the sample size improve the validity of the conclusions? *(1 mark)*

Glossary

Absolute fitness
   the ratio of frequencies of a particular genotype from one generation to the next

Active
   a process which requires energy

Active site
   the region of an enzyme molecule where the substrate binds

Anthropomorphism
   crediting animal behaviour with human emotions or qualities

Antigen
   a protein that may induce an immune response if it is foreign

Antigenic variation
   where parasites show great variety amongst different strains

Apoptosis
   programmed cell death

Autosome
   any chromosome that is not a sex chromosome

Bacteriophage
   a virus that targets a bacterial host

Caspase
   proteinases which destroy a cell

Causation
   when changing the independent variable causes the effect noted in the dependent variable

Centrifugation
   a process which uses centrifugal forces to separate components of a mixture

Chiasmata
   the place where two homologous chromosomes come into contact with one another

Chromatography
   a set of techniques which separates the components of a mixture

Civil engineering
   a profession that is involved in the design and manufacture of infrastructure to improve standards of living

Clonal selection
   lymphocytes become amplified, with some clones used in immediate defence and other clones acting as memory cells

Co-evolution
where a change in the traits of one species acts as a selection pressure on another species with which it frequently interacts

Colorimeter
a device that is used to measure the absorbance of a specific wavelength of light by a solution

Commensalism
an interrelationship between organisms of two different species in which one species benefits and the other neither benefits nor suffers

Competitive exclusion principle
when two species are in intense competition with one another and the weaker of the two species becomes locally extinct

Confidence intervals
a range within which the actual value should lie

Conformation
the spatial, or 3-D, arrangement of the atoms that make up the molecule

Correlation
when two variables seem to be connected

Crossing over
the process where homologous chromosomes swap genetic material

Cysticercosis
a parasitic tissue infection

Definitive (primary) host
the host where the parasite reaches sexual maturity

Electrophoresis
a process which applies an electric current across a gel to separate components of a mixture

Epidemiology
the study of the outbreak and spread of infectious diseases

Error bars
drawn on graphs to show the spread of data around the mean

Ethogram
chart on which observed animal behaviour is recorded

Ethology
study of animal behaviour

Eukaryotic
an organism which has a membrane bound nucleus

Evolution
>the change, over successive generations, in the proportion of individuals in a population differing in one or more inherited traits

Extended phenotype
>a theory whereby the parasite modifies the host's behaviour to increase its own transmission

Extracellular
>outside the cell

Fluorescence
>the emission of light of a different wavelength to that which was absorbed

Fundamental niche
>the niche that an organism occupies when there are no other species present competing for space or resources

Genetic drift
>the random change in how frequently a particular allele occurs within a small population

G-protein
>proteins which act as molecular switches - they allow signals from outside the cell to be transmitted inside (they are involved in signal transduction); their activity is regulated by their ability to bind and break down GTP to GDP - when GTP is bound they are 'on' and when GDP is bound they are 'off'

Gravid proglottid
>the segment of a tapeworm containing both male and female reproductive organs

Haemocytometer
>a device used to count cells

Hermaphrodite
>an organism with both male and female reproductive organs

Heterogametic
>dissimilar sex chromosomes, e.g. mammalian males where the Y chromosome is much smaller than the X chromosome

Homologous chromosomes
>chromosomes of the same size, same centromere position and which carry the same genes at the same gene loci

Hydrophilic
>(means 'water loving') a molecule of this type is electrically attracted to the polarity of the water molecules

Hydrophobic
>(means 'water hating') a molecule of this type is not electrically attracted to the polarity of water and is repelled away from the water molecule

GLOSSARY

Hyperpolarisation
a change in a cell's membrane potential that makes it more negative

Immune surveillance
white blood cells patrol the body, recognising and destroying foreign pathogens

Immunoassay
techniques which use antibodies linked to reporter enzymes to cause a colour change in the presence of a specific antigen

Independent assortment
takes place during meiosis I when homologous chromosomes pair up and line up along the equator - the final position of one pair is completely random relative to every other pair

Inflammatory response
injured or wounded areas become warm and red due to increased blood flow, bringing white cells for defence

Intermediate (secondary) host
the host that the parasite might require in order to complete its life cycle or as a means of transmission, thus making it a vector

Interphase
takes place at the start of meiosis when DNA replication occurs

Intracellular
inside the cell

Kinetochore
the point on a chromosome where the spindle fibre binds

K-selected
larger organisms that usually produce lower numbers of offspring, providing more extensive parental care and have longer life spans

Lekking
males display for females in a communal display area, then females choose a mate

Ligand
a substance which can bind to a protein

Line transect
line along which quadrats are placed or samples taken

Linked genes
genes that are on the same chromosome

Lymphocyte
a type of white blood cell which forms part of the immune system

Lysis
    the bursting of a host cell, releasing many virus particles

Malaria
    a serious human disease spread by mosquitoes that are infected with the malaria parasite

Mean
    this is the arithmetical average. Add all the values and divide by the number there are

Median
    this is the middle point from the range of values

Meiosis
    a special type of cell division where four haploid gametes are produced from one diploid gamete mother cell

Memory cells
    cloned lymphocytes that remain in the body to respond faster if the individual is exposed to the same antigen a second time

Mode
    this is the most common value in a set of data

Monogamy
    where animals form breeding pairs, thus one male to one female

Monomer
    a molecule that may bind chemically to other molecules to form a polymer

Mutation
    random change in DNA sequences within a population

Mutualism
    symbiotic relationship where both species benefit

Natural killer cells
    lymphocytes responsible for destroying abnormal cells

Natural selection
    non-random process whereby certain alleles occur more frequently within a population because they confer a selective advantage

Negative correlation
    as one variable is increased, the other variable decreases

Opsin
    a photoreceptor molecule found in the animal kingdom

Parasite
    an organism that gains food and shelter at the expense of the host

Parasite load
a measure of the number of parasite found within a host organism

Parthenogenesis
asexual reproduction whereby embryos develop from an unfertilised egg

Passive
a process which does not require energy

Phagocyte
white blood cell in non-specific defence, engulfing and destroying foreign antigens -may also present antigens to lymphocytes

Phagocytosis
non-specific defence where phagocytes engulf foreign antigens and digest them using digestive enzymes present in lysosomes

Phosphorylation
the addition of a phosphate group to a molecule

Photon
a particle representing a quantum of light

Placebo
a substance which has no therapeutic effect

Plagiarism
copying another person's work or views and passing these off as one's own

Plankton
microscopic organisms found living in fresh or salt water

Point count
sampling techniques used for birds

Polygamy
usually where one male has access to mate with several females

Polymer
a large molecule, or macromolecule, composed of many repeated subunits, known as monomers

Polypeptide
a single chain of amino acids

Positive correlation
as one variable increases, the connected variable increases too

Prosthetic group
a non-protein unit which is tightly bound to a protein that is necessary for its function

Qualitative data
: information that is observed and presented as a description

Quantitative data
: information that can be measured and is usually presented as a number

Random sampling
: individuals selected from the larger population must be chosen completely by chance

Range
: this is the set of values that the data falls into, thus the smallest value to the biggest value and everything in between

Realised niche
: the niche that an organism occupies when there is competition from other species

Recombinant
: the chromosome created when linked genes are separated during crossing over

Red Queen hypothesis
: a theory concerning co-evolution of a parasite and its host

Relative fitness
: the ratio of surviving offspring of one genotype compared with other genotypes

Resource partitioning
: where two species occupy different realised niches, allowing them to co-exist by compromising over resources

Retinal
: a light-sensitive molecule

Retrotransposon
: the piece of DNA that carries out reverse transcription before being inserted into a new site on the genome

Retrovirus
: a virus that contains RNA as its nucleic acid

Reverse transcriptase
: an enzyme used by retroviruses to synthesise DNA that can be inserted into the host cell's genome

r-selected
: smaller organisms that usually produce larger numbers of offspring, providing no parental care and having shorter life spans

Sanitation
: access to facilities for safely disposing of human waste such as urine and faeces

Satellite male
　　male that sneaks to gain access to females

Schistosomiasis
　　the human disease caused by schistomes reproducing sexually in the intestines

Selection bias
　　where individuals, groups or data are not selected randomly, therefore failing to provide a representative sample

Sessile
　　organism that is fixed in position - cannot move

Sexual dimorphism
　　physical difference between males and females of a species

Sexual selection
　　a non-random process, whereby certain alleles occur more frequently within a population because they are selected/preferred by one sex

Sneaker
　　male that gains access to mate with females without the more dominant male knowing

Standard curve
　　a graph which can be used to determine the concentration of an unknown solution

Standard deviation (SD)
　　shows how much variation or spread from the average/mean exists

Stratified sampling
　　individuals are randomly selected from sub-groups within a population

Summary
　　at the start of the report, where the main aims and findings are described briefly

Symbiosis
　　an interrelationship between organisms of two different species, whereby at least one species benefits

Symport
　　describes an integral membrane protein which is involved in simultaneously transporting two substances across the membrane in the same direction; in the case of glucose symport, glucose and sodium are simultaneously transported into cells

Systematic sampling
　　where samples may be taken at regular/pre-set intervals, e.g. every 2m along the transect

Taxonomy
　　grouping living organisms based on similarities or relatedness

Terrain
: the physical geography of the land

Transmission
: the spread of a parasite to a host

Vector
: the means of transmitting a parasite

Vegetative propagation/cloning
: a form of asexual reproduction that takes place in some plants, resulting in offspring that are genetically identical to the parent plant, e.g. bulbs and runners

Virulence
: the deleterious effect that the parasite has on the host

Waterborne
: transmitted by water

Answers to questions and activities for Cells and Proteins

Topic 1: Laboratory techniques for biologists

End of Topic 1 test (page 33)

Q1: Any two from:

- ensure that pregnant workers do not use it;
- wear gloves;
- wear a lab coat;
- have an established procedure for spills;
- maintain a clean working area before and after use;
- ensure that it is disposed of safely.

Q2: The likelihood of harm arising from exposure to a hazard.

Q3: Any from:

- a lab coat;
- safety glasses;
- a face shield.

Q4: To keep pH at a nearly constant value.

Q5: Colorimeter

Q6: 600 μg/ml

Q7: Immunoassay

Q8: (Thin sections of) tissue OR (individual) cells OR parts of organism.

Q9: Serum

Q10: To prevent contamination.

Q11: Any from:

- wiping down surfaces with disinfectant;
- working in a cell culture hood;
- washing hands before and after a procedure;
- using sterile equipment.

Q12: 2.4×10^5 OR 240000

Q13: Vital staining

Topic 2: Proteomics, protein structure, binding and conformational change

Extended response question: Structure of proteins (page 63)

Suggested marking scheme

Each line represents a point worth one mark. The concept may be expressed in other words. Words which are bracketed are not essential. Alternative answers are separated by a solidus (/); if both such answers are given, only a single mark is allocated. In checking the answer, the number of the point being allocated a mark should be written on the answer paper. A maximum of ten marks can be gained.

1. Primary structure of a protein is the sequence of amino acids in the polypeptide chain.
2. Amino acids are joined together by peptide bonds.
3. Secondary structure of a protein is stabilised by hydrogen bonds.
4. α-helix and β-sheet are two types of secondary structure.
5. α-helix is a spiral with the R groups sticking outwards.
6. β-sheet has parts of the chain running alongside each other forming a sheet.
7. The R groups sit above and below the sheet.
8. β-sheet can be anti-parallel or parallel.
9. Turns are a third type of secondary structure.
10. The polypeptide folds into a tertiary structure.
11. Folding at this level is stabilised by many different interactions between the R groups of the amino acids.
12. Any two from: hydrophobic regions, ionic bonds, hydrogen bonds, London dispersion forces, disulfide bridges.
13. A protein may include prosthetic (non-protein) parts.
14. For example, haem in haemoglobin.
15. Quaternary structure exists in proteins with two or more connected polypeptide subunits.

Extended response question: Enzyme activation (page 63)

Suggested marking scheme

Each line represents a point worth one mark. The concept may be expressed in other words. Words which are bracketed are not essential. Alternative answers are separated by a solidus (/); if both such answers are given, only a single mark is allocated. In checking the answer, the number of the point being allocated a mark should be written on the answer paper. A maximum of five marks can be gained.

1. Modulators regulate the activity of the enzyme.
2. Modulators bind at allosteric sites.
3. Causes a conformation/shape change in the enzyme.
4. Alters the affinity of the active site for the substrate.
5. Positive modulators increase the enzyme affinity for the substrate.
6. Negative modulators reduce the enzyme's affinity for the substrate.
7. Phosphorylation/addition of a phosphate or removal of a phophate/dephosphorylation can alter enzyme activity.
8. Kinase enzymes carry out phosphorylation/adds phosphate.
9. Phosphatase enzymes carry out dephosphorylation/remove a phosphate.
10. Some enzymes are activated by phosphorylation while others are inhibited.

End of Topic 2 test (page 63)

Q1: The entire set of proteins expressed by a genome.

Q2: Non-coding RNA genes

Q3: Metabolic activity OR cellular stress OR response to signalling molecules OR diseased versus healthy.

Q4: Golgi apparatus

Q5: Rough endoplasmic reticulum

Q6: Smooth endoplasmic reticulum

Q7: Along microtubules (to other membranes)

Q8: Addition of a carbohydrate group OR proteolytic cleavage OR phosphorylation/dephosphorylation/addition of a phosphate/removal of a phosphate.

Q9: Polar

Q10: Acidic/negatively charged

Q11: The order in which the amino acids are synthesised into the polypeptide.

Q12: Disulfide bridges/bonds

Q13: Proteolytic cleavage

Q14: α helix / α helices / alpha helix / alpha helices

Q15: A substance which can bind to a protein.

Q16: Allosteric sites

Q17: Changes in binding at one subunit alter the affinity of the remaining subunits.

Q18: b) Decrease in pH and increase in temperature.

Q19: Kinase

Q20: During a phosphorylation reaction the terminal phosphate of **ATP** is transferred to specific amino acid R groups. Addition of a phosphate group adds **negative** charges to a protein.

Q21: Phosphatase

Topic 3: Membrane proteins

Extended response question: The sodium potassium pump (page 79)

Suggested marking scheme

Each line represents a point worth one mark. The concept may be expressed in other words. Words which are bracketed are not essential. Alternative answers are separated by a solidus (/); if both such answers are given, only a single mark is allocated. In checking the answer, the number of the point being allocated a mark should be written on the answer paper. A maximum of ten marks can be gained.

1. The sodium potassium pump transports ions against a steep concentration gradient.
2. The sodium potassium pump requires energy from hydrolysis of ATP.
3. The maintenance of ion gradients by the pump accounts for a significant part of basal metabolic rate.
4. The protein has high affinity for sodium ions inside the cell/sodium ions from inside the cells bind.
5. The protein becomes phosphorylated by ATP...
6. ...which changes the conformation (of the pump)
7. The affinity for sodium ions decreases resulting in sodium being released outside of the cell.
8. Potassium ions from outside the cell bind to the sodium potassium pump.
9. Dephosphorylation occurs...
10. ...which changes the conformation (*award once only*).
11. Potassium ions are taken into the cell.
12. The affinity returns to the start.
13. Three sodium ions are transported out of the cell and two potassium ions are transported into the cell.

End of Topic 3 test (page 80)

Q1: 1, 2 and 4

Q2: 3

Q3: Hydrophobic interactions

Q4: Facilitated transport through transporter proteins is a **passive** process, meaning that it does not require energy.

Q5: Some transporter proteins require energy to bring about the necessary conformational change. In this case the transport is **active**.

Q6: Ligand-gated

Q7: Membrane potential

Q8: Voltage-gated

Q9: a) Potassium

Q10: b) Sodium

Q11: The sodium-potassium pump requires energy supplied by **ATP**.

Q12: Phosphorylation changes the conformation/shape of the sodium-potassium pump/protein *(1 mark)* this changes the pump's/protein's affinity for ions. *(1 mark)*

Q13: 36,000

Topic 4: Communication within multicellular organisms

Cell signalling: The action of testosterone (page 90)

Q1:

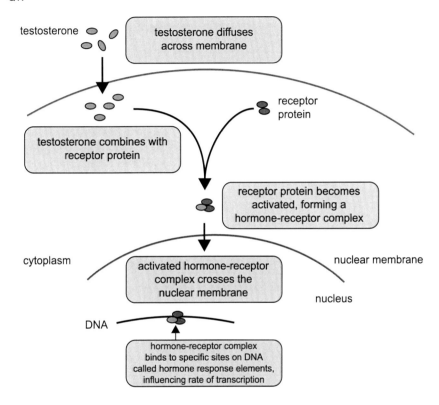

Extended response question: Cell signalling (page 109)

Suggested marking scheme

Each line represents a point worth one mark. The concept may be expressed in other words. Words which are bracketed are not essential. Alternative answers are separated by a solidus (/); if both such answers are given, only a single mark is allocated. In checking the answer, the number of the point being allocated a mark should be written on the answer paper. A maximum of ten marks can be gained.

1. Some signalling molecules/hormones are proteins/peptides/hydrophilic.
2. Hydrophilic/protein/peptide signalling molecules/hormones cannot cross the membrane.
3. The receptor for hydrophilic signals is in the membrane.
4. Binding changes the conformation of the receptor.
5. Receptors cause transduction/trigger cell response.
6. Example - insulin triggers recruitment of GLUT4/glucose transporters to the cell membrane.
7. Hydrophobic signals/steroid hormones can pass through the membrane.
8. The receptor for hydrophobic signals/steroid hormones is inside the cytosol/nucleus.
9. The receptors for hydrophobic signalling molecules are transcription factors.
10. The hormone-receptor complex binds to specific DNA sequences called hormone response elements (HREs).
11. Binding at these sites influences the rate of transcription.
12. Each steroid hormone affects the gene expression of many different genes.

ANSWERS: UNIT 1 TOPIC 4

End of Topic 4 test (page 110)

Q2: Cytosol/cytoplasm/nucleus

Q3: Hormone response element

Q4: On the cell surface OR on/in the membrane.

Q5: Steroid hormones/signalling molecules OR hydrophobic hormones/signalling molecules

Q6: Triggers recruitment of GLUT4 glucose transporters to the cell membrane of fat and muscle cells.

Q7: A state where there is no net flow of ions across the membrane.

Q8: Action potential

Q9:

1. Neurotransmitter binds to its receptor triggering the opening of ligand-gated ion channels.
2. Ion movement occurs and there is depolarisation of the plasma membrane.
3. Voltage-gated potassium channels open to allow potassium ions to move out of the cell.
4. This leads to a rapid and large change in the membrane potential.
5. The sodium channels become inactivated.
6. Voltage-gated sodium channels open and sodium ions enter the cell down their electrochemical gradient.

Q10: Sodium-potassium pump

Q11: Inability to generate an action potential/depolarise and therefore an inability to transmit nerve impulses.

Q12: Opsin

Q13: Retinal

Q14: Rod

Q15: G-protein/transducin

Q16: Phosphodiesterase/PDE

Topic 5: Protein control of cell division

Mitosis (page 118)

Q1:

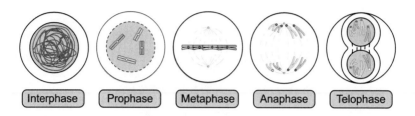

Extended response question: The cell cycle (page 127)

Suggested marking scheme

Each line represents a point worth one mark. The concept may be expressed in other words. Words which are bracketed are not essential. Alternative answers are separated by a solidus (/); if both such answers are given, only a single mark is allocated. In checking the answer, the number of the point being allocated a mark should be written on the answer paper. A maximum of ten marks can be gained.

1. Consists of interphase and mitotic (M) phase.
2. Interphase involves growth and DNA synthesis.
3. Interphase is G1, S, G2.
4. G1 and G2 are growth periods.
5. DNA replication occurs during S phase.
6. Mitotic phase involves mitosis and cytokinesis.
7. During prophase - DNA condenses into chromosomes each consisting of two sister chromatids OR nuclear membrane breaks down.
8. Also during prophase spindle microtubules (extend from the MTOC and) attach to chromosomes (via their kinetochores in the centromere region).
9. During metaphase chromosomes are aligned at the metaphase plate (equator of the spindle/cell).
10. During anaphase sister chromatids are separated, and the chromosomes are pulled to opposite poles.
11. During telophase the chromosomes decondense and nuclear membranes are formed around them.
12. During cytokinesis the cytoplasm is separated into two daughter cells.

Extended response question: Apoptosis (page 127)

Suggested marking scheme

Each line represents a point worth one mark. The concept may be expressed in other words. Words which are bracketed are not essential. Alternative answers are separated by a solidus (/); if both such answers are given, only a single mark is allocated. In checking the answer, the number of the point being allocated a mark should be written on the answer paper. A maximum of five marks can be gained.

1. Apoptosis is triggered by cell death signals that can be external or internal.

2. The production of death signal molecules from lymphocytes is an example of an external death signal.

3. External death signal molecules bind to a surface receptor protein (and trigger a protein cascade within the cytoplasm).

4. DNA damage is an example of an internal death signal.

5. An internal death signal resulting from DNA damage causes activation of p53 tumour-suppressor protein.

6. Both result in the activation of caspases (types of protease enzyme) that cause the destruction of the cell.

7. Cells may initiate apoptosis in the absence of growth factors.

End of Topic 5 test (page 128)

Q2: d) protein.

Q3: All eukaryotic cells have microtubules. These are hollow rods constructed of columns of a protein called **tubulin**.

Q4: In animal cells, microtubules radiate out from a region near the nucleus called the **centrosome/microtubule organising centre/MTOC**.

Q5: b) M, G1, S, G2

Q6: a) Anaphase

Q7: c) Prophase

Q8: b) Metaphase

Q9: d) Telophase

Q10: d) replicate its DNA.

Q11: Causes phosphorylation of proteins (that stimulate the cell cycle).

Q12: p53

Q13: Caspases

Q14: DNA damage OR absence of cell growth factors.

Q15: d) They cause cell proliferation resulting in tumour formation.

Topic 6: Cells and proteins test
Cells and proteins test (page 132)

Q1: Risk assessment

Q2: A colorimeter

Q3: 800

Q4: IEP or isoelectric point

Q5: Molecules are given an equally negative charge *(1 mark)* and denatured. *(1 mark)*

Q6: The antibody is linked to a chemical 'label'.

Q7: To estimate cell numbers in a liquid culture.

Q8: Serum

Q9: Eliminates unwanted microbial contaminants/prevent contamination.

Q10: Proteome

Q11: Hydrolases

Q12: Transport materials between membrane compartments.

Q13: Smooth endoplasmic reticulum or SER.

Q14: Peptide

Q15: Hydrogen

Q16: α-helix

Q17: Prosthetic group

Q18: Cooperativity

Q19: Catalyse the transfer of a phosphate group to other proteins.

Q20: Strong hydrophobic interactions hold **integral** proteins within the phospholipid bilayer.

Q21: The passive transport of substances across the membrane through specific transmembrane proteins.

Q22: Voltage-gated

Q23: Conformational change of membrane proteins in active transport requires energy from hydrolysis of **ATP**.

Q24: b) Three sodium ions out of the cell, two potassium ions into the cell.

Q25: The sodium gradient created by the sodium-potassium pump drives the active transport of glucose.

Q26: It changes conformation.

Q27: Hydrophobic signalling molecules can pass across the membrane.

Q28: Transcription factors

Q29: Loss of (insulin) receptor function.

Q30: It triggers recruitment of GLUT4/glucose transporters so can improve uptake of glucose to fat and muscle cells.

Q31: A wave of electrical excitation along a neuron's plasma membrane.

Q32: Depolarisation

Q33: Ligand-gated ion channels open.

Q34: Retina

Q35: Retinal

Q36: A single photoexcited rhodopsin activates **hundreds** of molecules of G-protein. Each activated G-protein activates one molecule of PDE. Each active PDE molecule breaks down **thousands** of cGMP molecules per second.

Q37: a) Cone cells do not function in low light intensity and contain different forms of opsins.

Q38: S

Q39: Degenerative

Q40: b) Metaphase

Q41: d) Telophase

Q42: c) Prophase

Q43: a) Anaphase

Q44: a) DNA replication to occur.

Q45: Spindle fibres

Q46: They phosphorylate proteins that regulate progression through the cycle.

Q47: Any from:
- stimulate DNA repair;
- stop/arrest the cell cycle;
- cause cell death/apoptosis.

Q48: DNA damage

Q49: Pilot study

Q50: To allow the chemicals/compounds to have an effect OR to allow apoptosis to begin.

Q51: Variability around the average/mean result.

Q52: Error bars do not overlap so differences (between rates of apoptosis) are significant/rates (of apoptosis) could not be the same.

Q53: The average percentage of apoptotic cells in the control is 7 and in the presence of ATO and ETME the average percentage of apoptotic cells 28, which is a difference of four times.

Q54: Western blot analysis shows that p53 protein expression is up-regulated/increased when cells are treated with the combination of ATO and ETME. *(1 mark)*
When a p53 inhibitor is present the apoptotic effect of ATO and ETME is reversed/reduced. *(1 mark)*

Q55: Caspase

Answers to questions and activities for Organisms and Evolution

Topic 1: Field techniques for biologists

Sampling of wild organisms: Questions (page 148)

Q1: 92

Q2: 96

Q3: Samples were taken every 3 m / at regular intervals.

Monitoring populations: Question (page 151)

Q4: 2160

Measuring and recording animal behaviour: Question (page 152)

Q5: B and D.
B. Offspring *begged* mother for food.
D. Offspring *smiled* at mother.

End of Topic 1 test (page 153)

Q6: Any two from:
- isolation;
- terrain;
- tidal changes;
- weather conditions.

Q7: Any from:
- camera traps;
- remote detection;
- scat sampling.

Q8: c) 10 pupils from each year group in a school are randomly selected and their pulse rate is taken.

Q9: Any two from:
- banding;
- hair clipping;
- painting;
- surgical implantation;
- tagging.

Q10: 500

Q11: Ethology

Q12: An ethogram

Q13: a) Anthropomorphism

Topic 2: Evolution

Drift and selection: Questions (page 157)

Q1: a) Genetic drift

Q2: b) Natural or sexual selection

Q3: b) Natural or sexual selection

Q4: a) Genetic drift

Fitness: Questions (page 158)

Q5: The ratio of frequencies of a particular genotype from one generation to the next is defined as **absolute** fitness.

Q6: The ratio of surviving offspring of one genotype compared with other genotypes is defined as **relative** fitness.

Co-evolution and the Red Queen: Questions (page 160)

Q7: 112.5 %

Q8: Number of nematodes in slug population 2 never increases higher than 25 in comparison to 450 in population 1.

Q9: Random mutation, resulting in resistance. Resistant individuals are more likely to reproduce so the resistance gene becomes more frequent in subsequent generations.

Q10: Nematode numbers would increase OR slug numbers would decrease.

End of Topic 2 test (page 162)

Q11: Genetic drift

Q12: Mutation

Q13: Relative

Q14: Red Queen

Topic 3: Variation and sexual reproduction

Costs and benefits of reproduction: Questions (page 168)

Q1: c) 4

Q2: c) Vegetative propagation

Q3: b) Parthenogenesis

Q4: a) Horizontal gene transfer

Meiosis forms variable gametes: Activity (page 171)

Q5:

1. Chromosomes undergo DNA replication (interphase).
2. Homologous chromosomes line up at the equator of the cell.
3. Homologous chromosomes touch at points called chiasmata.
4. Crossing over occurs at points called chiasmata.
5. Independent assortment occurs.
6. Two haploid cells are formed.
7. Chromatids are separated by spindle fibres.
8. Four haploid gametes are produced.

Meiosis forms variable gametes: Questions (page 172)

Q6: 6

Q7: 3

Q8: 8

Q9: LCPB or BPCL as per the following diagram:

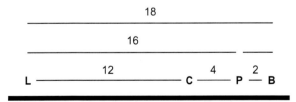

Sex determination: Questions (page 176)

Q10: 20 %

Q11: 10 %

Q12: 240

Q13: 540

Q14: 50 %

Q15: 50 %

Q16: 50 %

Extended response question: Meiosis (page 178)

Suggested marking scheme

Each line represents a point worth one mark. The concept may be expressed in other words. Words which are bracketed are not essential. Alternative answers are separated by a solidus (/); if both such answers are given, only a single mark is allocated. In checking the answer, the number of the point being allocated a mark should be written on the answer paper. A maximum of ten marks can be gained. A maximum of three marks can be gained from points 5-8.

1. Meiosis produces four haploid gametes...
2. ...from one diploid gamete mother cell.
3. Interphase is where DNA replications occurs.
4. During meiosis I, homologous chromosomes line up at the equator of the cell.
5. Homologous chromosomes are the same size and shape.
6. They carry the same genes at same gene loci...
7. ...but may carry different alleles / one from each parent.
8. They have their centromere at the same place.
9. During meiosis I, crossing over may occur...
10. ...at points called chiasmata.
11. This process shuffles sections of DNA between the homologous pairs, allowing the recombination of alleles to occur thus increasing variation.
12. Genes on the same chromosome are said to be linked.
13. There is a correlation between the distance between linked genes and their frequency of recombination / description or definition of chromosome mapping.
14. Independent assortment occurs as a result of meiosis I, with homologous chromosomes being separated irrespective of their maternal and paternal origin.
15. Homologous chromosomes are separated by spindle fibres.
16. This increases variation in the gametes.
17. During meiosis II, chromatids are pulled apart (and four haploid gametes are produced).

End of Topic 3 test (page 178)

Q17: c) Only half of each parent's genome is passed onto offspring.

Q18: b) increasing

Q19: b) This is common in plants.

Q20: A, B and C.
- A) Crossing over occurs at points called chiasmata.
- B) Independent assortment occurs.
- C) Homologous chromosomes are separated.

Q21: Any two from:
- same size;
- same centromere position;
- same genes at same gene loci;
- alleles may differ due to different parental origin.

Q22: Independent assortment and crossing over.

Q23: linked

Q24: Hermaphrodite

Q25: a) Heterogametic

Q26: a) one copy

Q27: b) two copies

Q28: Random inactivation of parts of the X chromosome OR half of the cells in any tissue will have a working copy of the gene.

Topic 4: Sex and behaviour

Sexual investment: Activity (page 185)

Q1:

Characteristic	r-selected population	K-selected population
Environment	Unstable	Stable
Lifespan	Short	Long
Number of offspring per reproductive episode	Many	Few
Number of reproductions in lifetime	Usually one	Often several
Size of offspring or eggs	Small	Large
Parental care	None	Often extensive

Courtship: Questions (page 191)

Q2: b) sexual dimorphism.

Q3: a) lekking.

Q4: c) sneaking behaviour.

Q5:

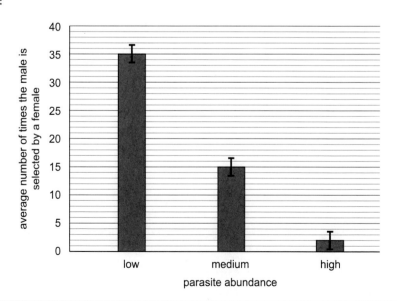

ANSWERS: UNIT 2 TOPIC 4

Q6: 57.1 %

Q7: The higher the parasite abundance the lower the mating success.

Q8: Low parasite abundance suggests that the male may have good parasite or disease resistant genes. These are 'honest' signals.

Extended response question: Courtship (page 193)

Suggested marking scheme

Each line represents a point worth one mark. The concept may be expressed in other words. Words which are bracketed are not essential. Alternative answers are separated by a solidus (/); if both such answers are given, only a single mark is allocated. In checking the answer, the number of the point being allocated a mark should be written on the answer paper. A maximum of ten marks can be gained.

Sexual diamorphism (maximum of 4 marks):

1. Sexual dimorphism is the physical difference between males and females of the same species.
2. Usually, males are more conspicuous than females / a suitable description.
3. Being inconspicuous mean that females can better protect their young due to camouflage.
4. Sexual dimorphism is a product of sexual selection.
5. Sexual dimorphism can be reversed in some species.

Male-male rivalry (maximum of 1 mark):

i. Males often use large size to out compete other smaller males for mates.
ii. Some males use weaponry to win females / suitable description of horns or antlers.

Sneakers (1 mark):

I. Smaller males may still be successful using sneaker / satellite behaviour.

Lekking (maximum of 4 marks):

a) Lekking is where males collect in a display area and present to females.
b) The display area is called a lek.
c) A suitable example, e.g. grouse, capercaillie.
d) Females assess male fitness...
e) ...and choose based on 'honest' signals...
f) ...such as low parasite burden.

End of Topic 4 test (page 194)

Q9: b) greater than

Q10: c) Production of many offspring, short life spans, no parental care.

Q11: Sexual dimorphism

Q12: b) To provide better camouflage when protecting their young.

Q13: Sneaking behaviour

Topic 5: Parasitism

Niche: Activity (page 202)

Q1:

Fundamental niche:	exists in the absence of interspecific competition.
Realised niche:	exists in the presence of interspecific competition.
Resource partitioning:	two different species compromise over resources to reduce competition.
Competitive exclusion principle:	competition between two species will see local extinction of the weaker.
Vector:	responsible for parasite transmission.
Definitive host:	where the parasite reaches sexual maturity.
Intermediate host:	required by some parasites to complete their life cycle.

Parasitic life cycles: Questions (page 205)

Q2: b) *Plasmodium*.

Q3: c) *Schistosoma*.

Q4: a) an arthropod tick.

Q5: Endoparasite

Q6: Human and tsetse fly.

Q7: The variant surface glycoproteins do not allow the immune system / white blood cells / antibodies to attach to fight or render the parasite harmless.

Q8: Antigenic variation means that one medicine / vaccine will not be suitable for all variants.

Parasitic life cycles: Activity (page 209)

Q9:

RNA	$\rightarrow$	DNA	$\rightarrow$	RNA
-	viral reverse transcriptase	-	host RNA polymerase	-

Transmission and virulence: Questions (page 210)

Q10: 157.1 %

Q11: b) 2011

Q12: c) 2012

Q13: Rabbits are an intermediate host so their greater numbers increased their role as a vector.
OR
The fox population increased due to food availability and parasite transmission increased due to overcrowding.

Defence against parasitic attack: Activity (page 213)

Q14:

1. Phagocytes move to the site of injury.
2. Plasma membrane of the phagocyte engulfs the parasite.
3. Parasite is brought into the phagocyte in a vacuole.
4. Lysosomes move towards the vacuole.
5. Lysosomes fuse with the vacuole releasing digestive enzymes.
6. The parasite is digested.

Defence against parasitic attack: Questions (page 215)

Q15:

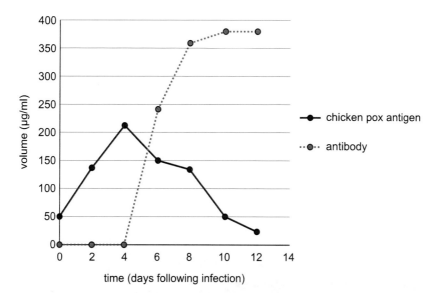

Q16: The volume of chicken pox antigen starts at 50µg/ml on day 0 (day of infection). This rises to a maximum volume of 220µg/ml 4 days following infection. The volume of chicken pox antigen then falls eventually to 20µg/ml 12 days following infection. *(units needed at least once)*

Q17: The volume of antibody remains at 0µg/ml from day 0 to 4 days following infection. This then increases to 240μg/ml 6 days following infection. This increases to 380μg/ml by 10 days following infection, where it remains steady to 12 days following infection. *(units needed at least once)*

Q18: Volume of chicken pox antigen falls quickly from 50μg/ml on day 0 to 0μg/ml 2 days following infection. This is because the volume of antibody increases rapidly to 240μg/ml 2 days following infection. By 6 days following infection the volume of antibody is at a maximum of 400μg/ml. This rapid secondary response is the result of clonal selection that occurred in the first chicken pox response and the immunological memory cells present in the individual.

Challenges in treatment and control: Questions (page 219)

Q19: Disagree - parasite abundance will be high, but country 3 will be higher. The population density may produce overcrowding and the sub-tropical climate will promote vector populations; however, with recent improvements in sanitation, parasite abundance should be falling. The hurricane season may cause sanitation problems and increase the chance of parasite abundance.

Q20: Agree - excellent sanitation, temperate climate and no overcrowding are perfect conditions for controlling parasites.

Q21: Disagree - country 3 will have much higher parasite abundance due to overcrowding and poor sanitation, compared with country 4 which has lower population density and increased education, promoting good sanitation. Climates are similar and may promote vectors, but country 4 is in a better state to cope.

Extended response question: Parasite niche (page 222)

Suggested marking scheme

Each line represents a point worth one mark. The concept may be expressed in other words. Words which are bracketed are not essential. Alternative answers are separated by a solidus (/); if both such answers are given, only a single mark is allocated. In checking the answer, the number of the point being allocated a mark should be written on the answer paper. A maximum of ten marks can be gained.

1. Parasites and hosts have a symbiotic relationship. . .
2. . . . that benefits the parasite at the expense of the host.
3. Parasite niche is a complex summary of tolerances and needs.
4. Parasites have narrow niche due to high host specificity.
5. Parasites are defined as degenerate due to loss of certain organs or structures.
6. Fundamental niche is the niche that exists in the absence of interspecific competition.
7. Realised niche is the niche that exists in the presence of interspecific competition.
8. Interspecific competition will result when two species have very similar niches.
9. Interspecific competition can lead to competitive exclusion principle, which is local extinction of one (the weaker) of the two competing species.

10. Resource partitioning allows two species to co-exist (by compromising).
11. Endoparasites live inside the host and ectoparasites live on the surface of the host.
12. A definitive host is the host where the parasite reaches sexual maturity.
13. An intermediate host may be a vector and is needed for the parasite to complete its life cycle.
14. Vectors have a role in parasite transmission from one host to another...
15. ...such as the mosquito spreading the malaria parasite / any other suitable example of a vector.

Extended response question: Immune responses (page 222)

Suggested marking scheme

Each line represents a point worth one mark. The concept may be expressed in other words. Words which are bracketed are not essential. Alternative answers are separated by a solidus (/); if both such answers are given, only a single mark is allocated. In checking the answer, the number of the point being allocated a mark should be written on the answer paper. A maximum of ten marks can be gained.

1. Defences can be non-specific / natural *and* specific / adaptive.
2. Physical barriers such as the skin prevent entry of parasites.
3. Chemical secretions such as mucus, tears, saliva and stomach acid.
4. Inflammatory response increases bloodflow and therefore phagocytes to site of injury or parasite.
5. Natural killer cells destroy *abnormal* cells.
6. Phagocytosis is where phagocytes engulf parasites into a vacuole/vesicle...
7. ...and digestive enzymes in lysosomes digest the parasite.
8. White cells carry out 'surveillance'.
9. Phagocytes display foreign antigens to lymphocytes.
10. A specific lymphocyte is produced in response to each foreign antigen.
11. Lymphocytes undergo mitosis so are amplified.
12. This is called clonal selection.
13. T cells / T lymphocytes target infected or damaged cells.
14. T cells induce apoptosis (cell death).
15. B cells / B lymphocytes produce specific antibodies to specific antigens.
16. Some cloned lymphocytes act as immunological memory cells.

Extended response question: The treatment and control of parasites (page 222)

Suggested marking scheme

Each line represents a point worth one mark. The concept may be expressed in other words. Words which are bracketed are not essential. Alternative answers are separated by a solidus (/); if both such answers are given, only a single mark is allocated. In checking the answer, the number of the point being allocated a mark should be written on the answer paper. A maximum of ten marks can be gained. Only two points from 10-13 can be used to gain marks.

1. Parasites are difficult to culture in the laboratory.
2. Many parasites have rapid evolution rates...
3. ...resulting in rapid antigen change.
4. This makes development of vaccines difficult.
5. Developing suitable drugs is also very difficult...
6. ...due to the metabolism of the parasite and their host being so similar...
7. ...since a drug intended to kill a parasite may also harm the host.
8. Control has been noted in areas with improved sanitation / civil engineering / sewage systems.
9. Where vectors are under control parasites are controlled too.
10. Control is particularly difficult in tropical climates / developing countries, ...
11. ...places where there is war / refugee camps, ...
12. ...areas facing natural disasters...
13. ...and overcrowded regions.
14. Improving parasite control reduces child mortality.
15. Where there are improvements, children are likely to thrive / increase in child growth, development and intelligence.
16. As children develop / become educated, they can continue improving control methods in these countries.

End of Topic 5 test (page 223)

Q22: a) benefit

Q23: b) detriment

Q24: a) narrow

Q25: a) high

Q26: Degenerate

Q27: Intermediate OR secondary host

Q28: Definitive OR primary host

Q29: Competitive exclusion principle

Q30: Resource partitioning

Q31: B and C

B. The nematode *Acsaris* inhabits the first 20% of the human intestine and the nematode *Strongyloides* inhabits the last 20% of the human intestine.
C. The mouse parasite *Heligmosomoides polygyrus* is suited to the acidic pH of the stomach, whereas the parasite *Litomosoides sigmodontis* has adapted to the more neutral pH of the small intestine.

Q32: b) Modification of host behaviour

Q33: Malaria and schistosomiasis.

Q34: Bacteria

Q35: Viruses

Q36: a) nm

Q37: 2, 4 and 5

- Lipid coat
- Protein coat
- RNA

Q38: 2 and 4

- They are composed of the host's materials.
- They help the virus remain undetected.

Q39: Reverse transcriptase

Q40: E, B, D, A, F, C

- Virus attaches to the surface of the host cell.
- Virus injects its DNA into the host cell.
- Virus interrupts the host cell's own metabolism, often entering the host's genome.
- Virus uses the host cell's machinery and raw materials to replicate its DNA and synthesise protein coat.
- The new DNA then enters the newly formed protein coats, thus producing many new viruses.
- These then leave the cell to infect new cells and the host cell undergoes lysis, bursting.

Q41:

Non-specific defence	Specific defence
Inflammatory response	Action by B cells
Mucus	Action by T cells
Phagocytosis	Clonal selection
Skin	Phagocytes present antigens

Q42: During phagocytosis, special organelles release digestive **enzymes** into the vacuole containing the parasite.

Q43: Lysosomes

Q44: Clonal selection

Q45: d) apoptosis of specific damaged cells.

Q46: a) Mimics host antigens.

Q47: b) will not

Q48: Herd or community immunity

Q49: Epidemiology

Q50: b) difficult

Q51: a) fast

Q52: b) difficult

Q53: b) similar

Q54: (Civil engineering leading to) good sanitation.

Q55: c) War

Q56: b) Tropical climates

Q57: a) Natural disasters

Q58: b) developing

Q59: a) decreases

Q60: b) increased

Topic 6: Organisms and evolution test

Organisms and evolution test (page 230)

Q1: Systematic

Q2: 280

Q3: Ethology

Q4: An ethogram is a list of all the behaviours shown by species. This can be used to produce time budgets for wild species. Measurements taken to produce a time budget include latency, and the **duration** and **frequency** of particular behaviours.

Q5: B and C
- Having fun.
- Making friends.

Q6: Genetic drift

Q7: Mutation

Q8: Neutral

Q9: a) Frequencies of a particular genotype from one generation to the next.

Q10: c) Surviving offspring of one genotype compared with other genotypes.

Q11: B, D, E
- Horizontal gene transfer.
- Short generations times.
- Warm environments.

Q12: co-evolution

Q13: Red Queen hypothesis

Q14: B and D
- Great variation may occur throughout offspring.
- Only half the population is able to produce offspring.

Q15: d) Vegetative propagation in plants is an example.

Q16: Parthenogenesis

Q17: c) Same genes at different gene loci.

Q18: a) Meiosis I

Q19: a) Meiosis I

Q20: b) 4

ANSWERS: UNIT 2 TOPIC 6

Q21: b) Haploid

Q22: a) different

Q23: Heterogametic

Q24: Linked

Q25: KNJML or LMJNK

Q26: c) A and D

Q27: Hermaphroditic

Q28: Environmental

Q29: c) Red eyed females and white eyed males.

Q30: A and B
- Inactivation is random.
- At least half of her cells will have a working copy of the gene.

Q31: d) Sperm: many; Eggs: contain an energy store.

Q32: Female

Q33: b) r-selected organisms produce many offspring, but provide no parental care.

Q34: Sexual dimorphism

Q35: B, C and D
- Large size.
- Sneaking.
- Use of weaponry.

Q36: Imprinting

Q37: b) Inconspicuous

Q38: d) Reversed sexual dimorphism

Q39: c) Lekking

Q40: a) Fitness

Q41: B, C and E
- Narrow niche.
- High host specificity.
- Often lack certain structures and organs.

Q42: b) degenerate.

Q43: a) competitive exclusion

Q44: c) fundamental

Q45: d) realised

Q46: e) resource partitioning.

Q47: Definitive (host)

Q48: Intermediate (host)

Q49: Virulence

Q50: a) Alteration of host foraging.

Q51: a) flatworms.

Q52: b) 2

Q53: b) viruses.

Q54: A, D, E

- Jointed legs.
- Segmented body.
- Exoskeleton.

Q55: A, C, E

- They are surrounded by a protein coat.
- They contain RNA or DNA.
- Some have a lipid membrane surround.

Q56: E, B, D, A, F, C

- Virus attaches to the surface of the host cell.
- Virus injects its DNA into the host cell.
- Virus interrupts the host cell's own metabolism, often entering the host's genome.
- Virus uses the host cell's machinery and raw materials to replicate its DNA and synthesise protein coat.
- The new DNA then enters the newly formed protein coats, thus producing many new viruses.
- These then leave the cell to infect new cells and the host cell undergoes lysis, bursting.

Q57: Reverse transcriptase

Q58: The **higher** the transmission rate, the greater the virulence.

Q59: Inflammatory (response)

Q60: C, A, B, D

- Phagocyte engulfs the pathogen into a vacuole.
- Lysosomes fuse with the vacuole.

- Pathogen is digested by enzymes.
- Product of pathogen digestion released into host cell cytoplasm.

Q61: b) antigens

Q62: a) antibodies.

Q63: c) apoptosis.

Q64: d) clonal selection.

Q65: Epidemiology

Q66: Herd immunity

Q67: b) suppress

Q68: b) increase

Q69: a) decrease

Q70: d) Developing countries have higher populations.

Answers to questions and activities for Investigative Biology

Topic 1: Scientific principles and process

Scientific cycle (page 249)

Q1:

1. Debating ideas and coming up a hypothesis to test
2. Researching others' work
3. Designing appropriate experiments
4. Observing and collecting data from experiments
5. Analysing data through comparing, interpreting and applying statistics
6. Evaluating results
7. Forming conclusions
8. Refining the original hypothesis

An investigation into whether athletes' body clocks affect competition performance (page 250)

Q2: To find out if athletes' body clocks affected competition performance.

Q3: Athletes' body clocks have no effect on competition performance.

Q4: **Competitors** - they will know that the timing of the event will not affect their performance.

Coaches - they don't need to worry about putting together different teams for different competition times.

Event holders - they won't be accused of being unfair due to the timing of their event.

Scientific ethics (page 253)

Q5:

Term	Definition
Refine	Ensuring competence in the experimental technique to reduce human error.
Replace	Using a different type of animal in the study.
Reduce	Using fewer animals in the study.

An investigation into whether athletes' body clocks affect competition performance (page 254)

Q6: Subject consent and subjects aware of the right to withdraw participation or data at any time.

Q7: With reference to another author, cited in the standard format (Parker et al, 2009).

End of Topic 1 test (page 255)

Q8: Null hypothesis

Q9: Peer review

Q10: a) Results can be negative, b) Results can be positive, d) Results will be verified independently and f) Results may cause a hypothesis to be rejected.

Q11: Scientific reports should be written in a manner that allows other scientists to *repeat* (*reproduce*) the experiment for verification or further work.

Q12: a) can reduce bias and b) has a standard format.

Q13: b) Reduce, replace and refine

Q14: c) Consent is required for all and d) Subjects can withdraw data at any time.

Topic 2: Experimentation

Controlled experiment 1 (page 268)

Q1: a) Positive

Q2: b) Negative

Controlled experiment 2 (page 268)

Q3: b) Negative

Q4: a) Positive

Extended response question: Pilot studies (page 278)

Suggested marking scheme

Each line represents a point worth one mark. The concept may be expressed in other words. Words which are bracketed are not essential. Alternative answers are separated by a solidus (/); if both such answers are given, only a single mark is allocated. In checking the answer, the number of the point being allocated a mark should be written on the answer paper. A maximum of three marks can be gained.

1. Plan procedures and assess validity / check techniques.
2. Evaluation / modification of experimental design.
3. Ensure an appropriate range of values for the independent variable.
4. Establish the number of repeat measurements required to give a representative value for each independent datum point.

End of Topic 2 test (page 282)

Q5:

Term	Definition
Precision	The closeness of repeated measurements to one another.
Accuracy	A measure of how close the data is to the actual true value.
Validity	Refers to the control of confounding variables.
Reliability	Refers to whether a procedure yields consistent results each time it is repeated.

Q6: b) Oxygen uptake

Q7: Multifactorial

Q8: Possible answers:

- some samples are a long way from best-fit line;
- there are small/no samples for some pouch volumes;
- there are some obvious discrepancies/inconsistencies, i.e. a wide range of frequencies at the same pouch volume.

Q9: Systematic sampling

Q10: This experiment was conducted *in vitro*. The *independent* variable (metal ions) is a **discrete** variable, while the **dependent** variable (enzyme activity) is a **continuous** variable.

Q11: Negative control

Q12: Any factor affecting the dependent variable that is not the independent variable.

Q13: Magnesium ions increase enzyme activity (by 36%) and all other metal ions decrease enzyme activity (ranging from 11% to 32% decrease).

Q14: There are no variation, error information or confidence intervals reported.

Topic 3: Critical evaluation of biological research

Evaluating background information (page 287)

Q1: Aim(s) and finding(s)

Q2: Relevant sources

Q3: Why this organism was chosen and any ethical considerations.

Evaluating experimental design (page 288)

Q4:

Term	Definition
Controls	Experiment set up to show that any effects are the result of the treatment
Controlled variables	Variables that may influence the dependent variable should remain constant throughout the experiment
Sample size	Large enough to allow a valid conclusion that the independent variable did have an effect on the dependent variable
Repetition	Carrying out the experiment again at a different time with new ingredients to determine reliability
Selection bias	Where individuals, groups or data are not selected randomly, therefore failing to provide a representative sample

Evaluating data analysis (page 291)

Q5:

Statistical term	Definition
Quantitative	Data that is measured
Qualitative	Data that is observed
Standard deviation	The spread or variation of data from the mean
Mean	Average
Mode	Most common value in a set of data
Range	Full set of data values
Error bar	Line drawn on a graph to show the spread of data around the mean

Dog foods experiment (page 292)

Q6:

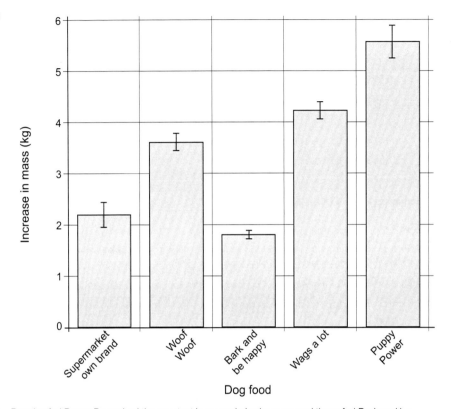

Q7: Puppies fed Puppy Power had the greatest increase in body mass, and those fed Bark and be Happy had the least.

Q8: No overlap between error bars / confidence intervals.

Q9: Bark and be Happy

Q10: Bark and be Happy due to the large variation in results, in comparison with Puppy Power where the range in data is much smaller.

Positive and negative correlations (page 294)

Q11: b) Negative correlation

Q12: a) Positive correlation

Q13: a) Positive correlation

Q14: a) Positive correlation

Q15: b) Negative correlation

Q16: b) Negative correlation

Q17: Yes, the two variables are connected. The correlation is negative; as age increases, milk production decreases.

End of Topic 3 test (page 298)

Q18: Aim

Q19: b) It should provide sufficient detail to allow repetition by other scientists, c) It should test the aims and hypotheses and d) It gives details of variables that should be controlled throughout.

Q20: Any four from:
- mean;
- mode;
- median;
- range;
- standard deviation;
- graphs;
- confidence intervals;
- error bars;
- t-test.

Q21: Sample size should be *large* and selected *randomly* to *avoid* selection bias.

Q22: c) As tail length increases, mating success in peacocks increases and d) As length of time as a gymnast increases, surface area of the cerebellum increases.

Q23: In order to reach a valid conclusion, procedures must be *evaluated*.

Topic 4: Investigative biology test

Investigative biology test (page 300)

Q1: b) To find out the effect of exercise intensity and duration on resting pulse rate.

Q2: c) Exercise intensity and duration will have no effect on resting pulse rate.

Q3: a) Ensures participants know they can withdraw at any time, c) Allows participants to assess any harm or risk that might be associated with the experiment and d) Allows participants to assess any benefits to society that may arise from the results of the investigation.

Q4: By using a cited reference (Wilson, 2014).

Q5: b) Whether participants participated in regular exercise or not

Q6: Participants received a health check prior to the investigation.

Q7: By ensuring each group had a similar mean age and age range.

Q8: Multifactorial

Q9: They provide results in the absence of the treatment.

Q10: By taking repeated readings of each individual's pulse rate.

Q11: Quantitative

Q12: Either of the following:

- Error bars/95% confidence intervals have a much wider range (± 14) in treatment 2 compared to only ± 3 in the treatment group.
- Error bars would overlap in treatment 2 when compared to treatment 1, but error bars in treatment one would not overlap with the control.

Q13: Negative correlation

Q14: 26.7%

Q15: Selection bias

Q16: Type of sugar

Q17: Grow *A. niger* with a growth medium containing sucrose.

Q18: A confounding variable is:

- a factor other than the independent variable that may affect dependent variable/results;
- any factor affecting the dependent variable that is not the independent variable.

Q19: The sugars tested will not result in citric acid production.

Q20: Glucose resulted in the highest citric acid production followed by fructose then lactose. No citric acid was produced from galactose.

Q21: There was no independent replication and the whole experiment was only carried out once.

Q22: Pilot study

Q23: Members of a population are selected at regular intervals.

Q24: Either of the following:
- it removes any psychological factors;
- it removes/reduces (researcher) bias.

Q25: Any one of the following:
- weight of participants;
- room temperature;
- type of reaction time test;
- time of day;
- timing of meals;
- time lapse between eating and conducting the test;
- type of food consumed;
- quantity of food consumed;
- liquid refreshments prior to the test;
- anxiety level/emotional factors.

Q26: Any two from the following:
- 500 mg caffeine results in a significant decrease in reaction time;
- 250 mg caffeine does not result in a significant decrease in reaction time;
- the placebo does not result in a significant decrease in reaction time.

Q27: Any one of the following:
- it would give a more representative sample;
- it would better represent the variation in the population;
- it would reduce the impact of uncontrollable confounding variables.